THE ELEMENTS
OF INVENTION

THE ELEMENTS OF INVENTION

Jeanne H. Simpson
EASTERN ILLINOIS UNIVERSITY

MACMILLAN PUBLISHING COMPANY
New York

Collier Macmillan Publishers
London

Editor: D. Anthony English
Production: Till & Till, Inc.
Production Manager: Aliza Greenblatt

This book was set in Electra by V&M Graphics, Inc., printed and bound
by Quinn-Woodbine.

Macmillan Publishing Company
866 Third Avenue, New York, New York 10022

Collier Macmillan Canada, Inc.

Library of Congress Cataloging-in-Publication Data

Simpson, Jeanne H.
    The elements of invention / Jeanne H. Simpson.
        p.   cm.
    ISBN 0–02–410621–6
    1. English language—Rhetoric.   2. Invention (Rhetoric)
I. Title.
PE1423.S56   1990
808'.042—dc20
                                                        89-12958
                                                             CIP

Printing: 1   2   3   4   5   6   7   Year: 0   1   2   3   4   5   6

Preface

When writers, especially student writers, announce that they can think of nothing to say, they miss the mark slightly. Usually, it isn't that there is an arid emptiness in the writer's head; rather the problem is a jumbled, inchoate mess of ideas, impressions, opinions, memories, and associations. The task of sorting through it all is daunting. This volume recognizes that "I can't think of anything to say" is really "Help me find a way to say all this and still make sense." Invention is offered here as the sorting mechanism, the collection of boxes and labels into which thoughts may go.

The invention systems presented are standard ones, all borrowed from other writers. These systems have worked many times for a variety of tasks. The selection is intended to offer student writers experience with the best known, most commonly used invention systems. While specific uses are suggested for many of the systems, I intend them only as suggestions, not as rules.

The headings anticipate questions student writers might ask about invention. Systems are often presented as pairs in response to these questions; a pair of invention systems seems to produce a larger, richer set of responses than a single system does. Besides, taking two runs at a subject encourages more drafting and more revision than inexperienced writers usually attempt.

Invention has the virtue of existing at a distance from editing, so that it provides writers with a chance to experiment, take risks, and practice without penalty. It should be fun. While there are assignments offered here to give opportunity for trying various systems, the spirit of the book is one of playfulness and experimentation. Writers need a chance to fool around with words and ideas without great cost (grades). I hope students will take invention seriously but not solemnly.

I wish to thank Robert Funk for suggesting me for this project and Tony English, my editor, for his patience and calm assistance. I appreciate the many helpful suggestions provided by the reviewers. To Betty McMahan, Susie Day, Bob Funk, and Jan Youga, my gratitude for being examples to follow. And above all, thanks to my family for their encouragement and support.

Jeanne Simpson

Contents

Chapter 4

Chapter 5

Chapter 6

Chapter 7

THE ELEMENTS
OF INVENTION

Chapter 1

What is invention and why do I need to know about it?

Invention techniques help you to find something to say. This book describes a variety of methods for that task. The methods work for a range of situations, such as not having any idea of what to write, or knowing perfectly well what you want to say but having no luck getting it out on a page.

It may seem strange to associate the word invention with writing. Most people consider invention as having to do with new gadgets, Thomas Edison, and the like. However, if you consider invention as finding new ways of seeing ideas, new ways of arranging them, and new ways of developing them, then the word fits both mousetrap building and essay writing.

Invention means doing some structured mental work. Gadget inventors don't proceed randomly; they study a problem systematically and solve it in steps. You can do the same thing as you write by using an invention method.

Invention indirectly improves your style and your control of editorial matters. It will make your stylistic and editorial choices easier because you will know more about what you intend to say.

Are there any rules I have to memorize?

There are suggestions and guidelines for using invention. There are not any hard and fast rules and certainly nothing as demanding as the rules of English grammar. Invention relies more on what French drivers refer to as **System D**. System D means *"Do Whatever Works."* In Parisian traffic, that may include driving on the sidewalk. In invention, unconventional thinking is not only acceptable, it is one of the goals you are aiming for.

How can I get the most out of an invention system?

1. **Try to do all the steps in an invention system.** If you don't, the world won't end, but you might miss a good opportunity to explore an idea.
2. **Ignore redundancies.** Most invention systems will have you repeating yourself a bit. It's all right to do that. In fact, you may get a better grip on a slippery idea the second time you write about it.
3. **Write as much as you can in response to any step in a system.** Don't worry about the beauty of your writing; the main thing is to get your ideas down on paper where you can see them and work with them.
4. **Forget about grammar, punctuation, and spelling for now.** Invention is the one part of writing that does not involve worrying about commas and spelling or even complete sentences. Just dump the ideas down however you can. You can clean them up later. Arrows, abbreviations, shorthand, fragments, lists, circles, and underlining are all perfectly acceptable. Invention is for you to work from. It is **not** a perfect outline of your paper.
5. **Keep going, even if you are drawing a blank.** Make the pen, pencil, or cursor move no matter what. You can write "I can't think of a blessed thing to say" or "This system isn't working" or "zzzzzzzzz," as long as you keep your hand moving. Eventually, something will occur to you. Invention is a good way to break writer's block. Besides, you can enjoy a feeling of accomplishment when you've filled three or four pages with your invention notes.
6. **Reward yourself when you've finished a good workout**

with an invention system. Take a break, perhaps have a snack. A mental "cool down" period is often as important as a physical one; use it to help you gain some perspective on what you have written so far.

Following these guidelines is the most important part of invention. Doing invention the right way just means doing it consistently and with commitment. **The only wrong answer in invention is a blank page.**

Chapter 2

Is there something easy I can try first?

Cubing

One of the easiest invention methods is called **cubing**. Developed by Elizabeth and Gregory Cowan, it gets its name from the six steps involved in the system, equivalent to the six sides of a cube.*

Cubing is a handy invention method because you can memorize it and use it without having any special book or paper in front of you. You can use it when you are up against a tight deadline. Cubing can help you answer essay questions. It can help you retrieve ideas when you have left everything at school.

Cubing is also an excellent study technique. Once you've cubed on a subject, such as photosynthesis or the importance of railroads in the opening of the West, you will have a fair amount of control over your information and you should be prepared to answer most typical test questions.

Remember to **keep writing** and to ignore editorial issues for now.

HOW TO CUBE

Plan on writing nonstop for about three to five minutes for each part of the cube. Don't worry about doing it "the right

*Cowan, Elizabeth and Gregory Cowan. *Writing*. New York: John Wiley & Sons, 1980, pp. 25–26.

way." Remember System D — if it seems to be working, it **is** working. Feel free to make something up, as long as you remember that it is made up and not a fact.

Side 1: Describe

Write a description of whatever you are writing about. Your description can include color, shape, age, dimensions, location, sequence, contents, substances, style of construction, origin, maker, nationality, and so on.

Side 2: Compare

Write comparisons between your subject and any other things that come to mind. You can write an extended comparison and explain the similarities that you see. Or you can write a series of one-sentence comparisons, "X is like Y."

Side 3: Associate

Write about the events, ideas, concepts, persons, times, and places you associate with your subject. The reasons for associating them need not be explained unless you want to. The associations may be personal ones connected to your own experiences, or they may be abstract associations based on cause and effect relationships of some kind.

Side 4: Analyze

Write about the parts of your subject, how they relate to each other. Write about the origin of your subject and then about any outcomes or conclusions.

Side 5: Apply

Write about how your subject might be used. Can you learn something from it? Can you use it as a tool? How?

Side 6: Argue for or against

Pick a position with regard to your subject and defend that position.

EXAMPLE OF CUBING

The following example of cubing was written by a student trying cubing out for the first time. The student was given a piece of black licorice candy to write about. It may seem to be a trivial subject, but the student was able to say quite a bit about the licorice. Imagine the results if the subject were larger or more serious. Practicing with something small and concrete like a piece of candy is a good way to get the feel of an invention technique. The more you practice, the more comfortable you will become with inventing.

MY PIECE OF CANDY

[The student felt obligated to put a title at the top of the page. It didn't hurt anything to do that.]

Describe it.

This piece of black licorice candy is shiny and about 6 inches long by 1/2 inch wide and twisted. I can't think of anything else to say about it. I hope I'm doing this right. XXXXXXXXXX I am SO hot, I hope this candy doesn't melt all over my desk its already getting sticky from where I've handled it. XXXXXXXXXXXXXX. Uh oh, she saw me staring off into space I better think of something to say. The candy smells like licorice. It tasted like licorice which I hate and now I have this awful taste in my mouth for the rest of the hour yuk. Maybe I

Compare it.

Black licorice is like road tar. Especially today. It is SO hot in here. Black licorice is like dark glue. My candy is like my little sister; it sticks too much and annoys me. Here I am again, struggling to find something to say. What can I say? Licorice is like spoiled mouthwash. Licorice is like getting coal in your stocking for Christmas. Hey, not bad. Licorice is like

Associate it.

The Red and White Store near my elementary school had a candy counter and I used to go there during lunch. I couldn't go after school because I had to catch the bus home. They had sour grape bubble gum balls and jawbreakers and red and black licorice sticks. A whole bunch of us would go over and buy little white bags of candy. I had never eaten licorice and one of my friends told me it was delicious so I bought a sack ful. I waited until we got back to the playground to try it while waiting my turn at jacks. I took a big slurp on the first stick. I spit it out. I never spit my mother says spitting is the rudest thing there is but I did it anyway. I hated that licorice and I was furious that I had spent all my candy money on it. I had to wait a whole week to buy any more. After that I always bought sour grape bubble gum balls. My friend said that I

Analyze it.

I have no idea what licorice is made out of. Probably developed by a dentist because it can pull out a perfectly good filling as fast as anything I ever saw. A greedy dentist. The flavor comes from some seed. I think they make some drink out of it that's really popular in ------XXXXXXX I can't remember. It's probably made by demented elves in a dark cave somewhere, ones who won't be in TV commercials. Who hate kids and lay traps for them in candy stores. I bet it has monodiglicerides (sp?) and sodium caseinate and fake sugar and caffeine and nicotine and everything else that causes cancer in lab rats in it. I've said enough about THAT XXXXXXXXXXXXXXXXXXXXXXXXXXXXXXXXXX I hope she calls time soonXXXX

Apply it.

Poisining little sisters. Torturing students. Patching chuck holes. Oh lord the person next to me is drinking her

Dr.Pepper through her candy like a straw. Barf! I think I'll move to another seat next class.APPLY I have no use for this stuff but people who like it can have it. I won't ever buy it for Trick or Treat unless I'm on a diet and want to be sure I give all the candy away and don't try any myself which doesn't sound fun at all. Besides it doesn't come in individual wrapping. Should I put that under description? Who CARES?!? I hope we're nearly through with this because I am sick of this dumb candy and its smell and all.XXXXX

Argue for or against it.

That's easy. I am against this stuff. I think I have already made that clear but she said we would repeat ourselves alot. We could use it as chemical warfare, send it to the Russians and Cuba in big barrelsXXXXI could use it to control Susie when she's a nuclear brat acceptXXOr is it except I can't remember anyway she'd probably like it   Oh good she's looking at her watch so we must be nearly done   Have I argued or just applied some more? It is bad for the teeth and the environment so we should abolish licorice. There.PLEASE PLEEEZE call time my hand hurts

Comments

Note that this writer begins to say more and more as the cubing progresses. The segment associating the licorice is the most coherent and detailed section, so that part might be the nucleus of a short paper about a personal experience. At the same time, the student reveals an interesting and humorous voice that is not really apparent in the first segment. By the time the cubing is over, the writer's style has settled into a conversational, amusing tone that could transfer well into a paper. The writer also begins to think about what she is saying that she likes and may want to use. She enters into the spirit of invention by making up fantasy about the licorice when she knows she has no real information. She also indicates when she might need to look something up — the country where a licorice-flavored drink is popular.

At the beginning of the exercise, the writer is clearly unsure and uncomfortable. She is trying to play by the rules. She never entirely lets go of her editor, worrying about spelling at one point. Still, the fragments and misspellings go unchanged, indicating that she succeeded in focusing mostly on the subject of the candy. The final sections are not as well developed as the association section, suggesting that by then she had said the most important things. There are, however, in all of the later sections, some interesting ideas that she can probably use.

How do I develop my invention writing into a paper to be graded?

The example of cubing is interesting but obviously is not polished enough to be called a paper. The writer can lift whole sentences and phrases out of it to use though, making the work of creating a real essay much easier.

One way of making invention work is to go back and underline or mark the sentences or words or phrases you like or that you would like to develop further. Then try numbering the ideas in an order you think might work for a paper.

FOR PRACTICE

Try marking the ideas in the cubing example that you would use for a short essay. Number them in the order you would like. Read the numbered ideas out loud to hear what a rough draft of the essay would sound like.

Will this work for a real paper?

The licorice invention is just an exercise. It is unlikely that you will be asked to write an essay about a piece of candy; the idea is to provide evidence that invention systems do generate a surprising amount of material even on trivial subjects. When you must write on a more demanding subject, cubing is still a useful approach, especially for writers who aren't accustomed to doing invention.

ASSIGNMENT

*Write about a time when you had to react quickly to a threat.
Explain what the threat was or why it seemed to be a threat.
Describe your reaction. Explain what you learned from the experience. (This may be a safety precaution, a new way of seeing
another person, or an attitude you developed.)*

Before cubing for this assignment, study the imperative
verbs in it; they will tell you what you need to do to fulfill
the assignment. They include "write," "explain" (twice), and
"describe." You will have to write, of course. "Explain" can
mean a number of things, many of which occur in cubing.
When you compare, analyze, apply, or associate, you are
explaining. And describe is one of the steps in cubing.
Cubing, then, will help you satisfy the requirements of the
assignment.

Example of Cubing for Assignment

Describe

*I had had my driver's license for about 6 mo. when my
dad asked me to go to the hardware store for him and buy a
dozen screws to match the one he gave me. I held the screw
in my hand and drove off. About halfway to the store I
dropped the screw between the seat and the door. I looked
down to get it and when I did I drove straight into two
parked cars. I was entering a curve and so I drove straight
instead of turning slightly. I wore a seatbelt which was a good
thing because the force of it threw me forward hard enough to
make my chin and jaw hit the steering wheel. I bit through
my lower lip and bled*

Compare

*Hitting those cars was like jumping off the high dive
except I didn't expect the impact. My reaction was like when
I flunked algebra once — I knew my dad would kill me. It was*

as dumb to stop looking as   I can't think of anything but I
felt so STUPID after all the times Mr. Gunn told me to keep
my eyes on the road in drivers' ed. When I went to school I
felt like you feel in those dreams when everybody is dressed
up and you come to work in your underwear because I was
sure everybody was staring at me and thinking she sure is
dumb

Associate

"Keep your eyes on the road
And your hands upon the wheel"
Even Jim Morrison knew better. But I remember finding out
that my best friend did about the same thing but he didn't
wear his seatbelt and he broke his jaw and made a mess of
his lower teeth. I got off easy. Going to court to face a judge
scared me out of my wits even though he was nice

Analyze

I dropped the screw, which was the real problem. I should
have put it in my purse or pocket or something. Holding it
was stupid. I took my eyes off the road and didn't take my
foot off the accelerator. The weirdest thing was my reaction. I
sat there for a long time with the hood up in a mess in front
of me so I couldn't see what I had done. I don't think I ever
saw the other cars before an ambulance came to take me to
the hospital. Anyway I sat there thinking that if I just wished
hard enough it would disappear and I would be driving to
the hardware store no problem like I was a kid believing in
fairies to save Tinkerbell.

Apply

I certainly believe in seatbelts because a little scar on my
chin is nothing to the full bridge my friend is stuck with. I
have a terror of tickets and cops and judges and probably still
my father. I am the carefullest driver you ever saw. There
must be something in the growth hormones or something that
makes people have wrecks when theyre 17 because everybody I
know did. I don't want to write some preachy junk about
being a safe driver, but that is

Argue

> *Argue in favor of a wreck. Right. I can't argue anything here, there isn't anything to argue about Shall I just write XXXXX for five minutes or what?*

Comments

This example of cubing offers a number of possibilities for developing an essay describing a response to a threat. The writer needs to go back to the assignment to determine the required elements in the paper and then see where in the cubing exercise material occurs which corresponds to those elements.

The assignment requires a description of the threat. The writer here does not spend much time actually describing a threat. Sentences that seem to refer directly to a threat include

1. ". . . I drove straight into two parked cars."
2. ". . . the force of it threw me forward hard enough to make my chin and jaw hit the steering wheel. I bit through my lower lip and bled"
3. "I knew my dad would kill me."
4. "When I went to school I felt like you feel in those dreams when everybody is dressed up and you come to work in your underwear because I was sure everybody was staring at me and thinking she sure is dumb"
5. "Going to court to face a judge scared me out of my wits even though he was nice"
6. "I have a terror of tickets and cops and judges and probably still my father."

There are actually several threats referred to in these sentences. The threat of injury in the collision is one. But the writer also refers to the threat of embarrassment at school, the threat of parental displeasure, and the threat of going before the judge. The writer will need to clarify the relationships between all these threats so her audience will understand how they fit together and which ones are the most important.

The assignment also requires a description of the writer's response to the threat. Several vivid passages from the cubing describe responses.

1. "Hitting those cars was like jumping off the high dive except I didn't expect the impact."

2. "My reaction was like when I flunked algebra once — I knew my dad would kill me."
3. ". . . I felt so STUPID after all the times Mr. Gunn told me to keep my eyes on the road in drivers' ed."
4. "Anyway I sat there thinking that if I just wished hard enough it would disappear and I would be driving to the hardware store no problem like I was a kid believing in fairies to save Tinkerbell."

All these statements have strength, but in their present order, they do not reflect the sequence of events. The writer later re-arranged them so that the hitting analogy comes first, the wishing sentence second, the algebra analogy third, and the description of feeling stupid fourth.

Finally, the assignment calls for analysis of what the writer learned from the experience. The writer has three sentences that address that requirement.

1. "I certainly believe in seatbelts because a little scar on my chin is nothing to the full bridge my friend is stuck with."
2. "I have a terror of tickets and cops and judges and probably still my father."
3. "I am the carefullest driver you ever saw."

As in the case of the list of threats, here the writer needs to clarify the relationships between the things she learned from this experience. She might, for example, rank order them so that her effort to be a careful driver is the most important, her belief in safety devices such as seatbelts next most important, and her fear of receiving a ticket least important.

Obviously, this writer still doesn't have a fully developed essay. What she **does** have is a rough outline and a clearer sense of what problems are left to be solved as she drafts. She knows that she can follow the outline implied by the assignment: threat, reaction, lesson. She knows some of the specifics that she will include under each of those main headings. She will need to clarify relationships and develop detail. However, in developing those details, she already has some good images to use, such as the one about wishing the wreck would disappear.

What should I do when I have trouble sorting out my ideas?

In order to clarify material, the writer can try another invention technique called **looping**. It was invented by Peter Elbow.* Looping is particularly helpful when you want to explore more about an idea you think might be important. It can also be used to sharpen a thesis idea, so that you say what you really have in mind.

HOW TO DO LOOPING

As in cubing, plan to write nonstop for three to five minutes during each section. Remember not to worry about editorial matters. Just keep the pen, pencil, or cursor moving.

Looping is less structured than cubing. There are no specific steps involved. You should try at least three loops, however, in order to get useful results.

In the first loop write about your subject for several minutes. At the end of the time, stop and read what you have written. Under your first loop, write a sentence that contains a summary of your main idea. This is called a **center of gravity sentence**. It is designed to capture the essence of your ideas into a single, clear statement. Take a little time so that you are satisfied with your center of gravity sentence.

Your second loop should be a response to the center of gravity sentence. You may expand on its ideas, or you may want to examine any contradictions or problems you see in your first sentence. At the end of your second loop, prepare another center of gravity sentence.

The third loop works the same way as the first and second. At the end of it, write another center of gravity sentence. By this time, you should be getting both repetition and some refining of your thoughts.

---

*Elbow, Peter. *Writing without Teachers.* New York: Oxford UP, 1975.

EXAMPLE OF LOOPING

The threat paper needed more work, so the writer used looping to sort out the section on the nature of the threat. Notice how she searches for a way to present the threat so that her readers will understand its complexity.

1st Loop

I never really saw any threat. By the time I understood anything the wreck was over. All I saw was the hood up in front of me. I never really saw the mess I had made, but my dad certainly explained it to me when he filed the insurance claim. He explained again when he told me our insurance on me was cancelled. All I knew was that I felt this terrific jolt and heard a huge noise of crashing metal. I don't remember ever seeing the cars I hit.

CENTER OF GRAVITY SENTENCE
Maybe the whole point of the experience is that I should have been paying attention so I would see the threat in time to prevent hitting the cars.

2nd Loop

If I had seen the cars, I wouldn't have hit them. So in this case the threat maybe wasn't curves in the road or parked cars but a threat I think about now, which is not paying full attention to driving. I see people all the time yelling at kids or combing their hair and that guy I passed who was reading a book while he drove I sometimes space out too. I come to and realize I've been thinking about something and ten miles have passed without my noticing anything. Scary. But I don't take my eyes off the road

CENTER OF GRAVITY SENTENCE
Not paying attention to driving is a threat a lot of people forget about.

3rd Loop

When you do something everyday it is easy to stop paying attention to it. You don't think about a car being able to do that much damage. And the insurance company is just an envelope and a bill a few times a year. So you don't think about any connection between that and the driving. When it really happens to you, it's so completely unexpected that you can't believe it because you just don't think about it happening. It isn't even being too confident. It is just not noticing the possibilities of what you're dealing with.

CENTER OF GRAVITY SENTENCE

I forgot about the responsibility of driving a car and how much potential for danger is involved so that the threat to me was caused by my own forgetfulness.

Comments

The writer has found a way to define the threat so that all of the ideas developed in the cubing exercise can be organized. While the last center of gravity sentence is not especially graceful, it does have the makings for a controlling idea for the essay.

The writer now has enough to begin a first draft by pasting together the pieces of invention developed so far.

EXAMPLE OF DRAFTING FROM INVENTION

I forgot about the responsibility of driving a car and how much potential for danger is involved so that the threat to me was caused by my own forgetfulness. Not paying attention to driving is a threat a lot of people forget about. When you do something every day it is easy to stop paying attention to it.

I had had my driver's license for about 6 months when my dad asked me to go to the hardware store for him and buy a dozen screws to match the one he gave me. I held the screw in my hand and drove off. About halfway to the store I dropped the screw between the seat and the door. I looked down to get it and when I did I drove straight into two parked cars. I was entering a curve and so I drove straight instead of turning slightly.

Hitting those cars was like jumping off the high dive except I didn't expect the impact. My reaction was like when I flunked algebra once — I knew my dad would kill me. I felt so STUPID after all the times Mr. Gunn told me to keep my eyes on the road in drivers' ed. When I went to school after the wreck I felt like you feel in those dreams when everybody is dressed up and you come to work in your underwear because I was sure everybody was staring at me and thinking she sure is dumb. "Keep your eyes on the road and your hands upon the wheel." Every time I hear Jim Morrison's song, I think about that wreck.

Dropping the screw was the real problem. I should have put it in my purse. Holding it was dumb. I took my eyes off the road and didn't take my foot off the accelerator. After I hit the cars, the hood flew up and I couldn't see anything. I don't think I ever saw those cars. But I knew what happened. The weirdest thing was my reaction. I sat there for a long time with the hood up in a mess in front of me. I sat there thinking that if I just wished hard enough it would disappear and I would be driving to the hardware store no problem like I was a kid believing in fairies to save Tinkerbell.

I got off easy. I had to go to court to face a judge and that scared me out of my wits even though he was nice. Later I found out that a college friend had done about the same thing, but he didn't wear his seatbelt. He broke his jaw and made a mess of his lower teeth. I wore a seatbelt which was a good thing because the force of the wreck threw me forward hard enough to make my chin and jaw hit the steering wheel. I bit through my lower lip and bled a lot. I certainly believe in seatbelts because a little scar on my chin is nothing to the full bridge my friend is stuck with. I have a terror of tickets and cops and judges and probably still my father. I am a really careful driver now.

Comments

This draft needs to be expanded a little and to have the errors in punctuation and so on cleaned up. It is, however, a good working draft that meets all the requirements of the original assignment.

ASSIGNMENT

Revise this draft into a more finished paper. Discuss your revisions in class.

Invention is the step before drafting. Drafting involves putting ideas into a preliminary order so that you can tidy them up. Invention involves getting those ideas in the first place. Cubing and looping are two easy ways to find ideas, and they work well as a combination.

Cubing is an example of formal invention, or invention that takes a specific form, such as answering a series of questions or making guided responses. Looping, on the other hand, is an example of informal invention, or invention that is less structured and that does not require you to answer any specific questions.

Should I use formal or informal invention?

Some writers seem to have more success with formal invention, some with informal methods. Probably you will find that you have the greatest success if you use a combination of methods, similar to the preceding cubing and looping combination. Different parts of the writing process may call for different invention methods. The more methods you know, the easier it will be for you to select the one you need. Invention techniques are tools; having the right one will make your job much easier.

Chapter 3

In Chapter 2, you encountered an informal invention technique called looping. Although it doesn't involve answering a set of questions, as occurs in formal systems, looping is still somewhat structured because you take a series of steps toward a specific focus for your ideas.

Peter Elbow developed another invention method that is even less structured than looping. It is called **freewriting**.* Freewriting means just what it sounds like: writing without attention to rules or specific focus. You write for five, ten, or fifteen minutes, nonstop. As before, you don't pay attention to spelling, grammar, punctuation, complete sentences. Probably you will write more fluently than you expect when you freewrite, but correctness is not an issue here.

When should I use freewriting?

The purpose of freewriting is to get you warmed up, both physically and mentally. If you begin just writing about whatever comes to mind, you may not be on the subject you'll eventually cover, but you **will** be thinking about the act of writing itself.

Freewriting, like exercise, works best if you do it regularly. If you spend four or five minutes freewriting either before or after a class, it will help you get your thoughts channeled into the material covered. If you spend a little time freewriting

*Elbow, Peter. *Writing without Teachers.* New York: Oxford UP, 1975.

before you begin a paper, you will find some ideas beginning to gel. E. M. Forster once said, "How do I know what I think until I see what I say?" Freewriting gives you a chance to know what you think.

One of the pleasant side effects of freewriting is that the more you do it, the better writer you will become. The more writing you do, the more you will learn about yourself as a writer. Freewriting offers a low-risk opportunity for practice.

USE FREEWRITING FOR:

Warming up to write papers
Gathering thoughts before class
No-risk practice writing
Understanding what you think

In the following example of freewriting, the student is reacting to a column by Mike Royko. This response is not an essay or even the beginning of an essay. It represents Jon's raw thoughts, an immediate reaction. In the column, Royko referred to William Bennett's idea that children should turn in their drug-using friends and parents. The example shows a student seeing what he thinks.

FIRST EXAMPLE OF FREEWRITING

RESPONSE TO ROYKO/BENNETT

Protection is such an ugly term. It brings to mind images of secret service men rushing around the president, crash dummies being saved from bashing through windshields by seatbelts, and twelve-packs of condoms. Trying to protect someone from themselves is a very distinct way of denying that the person is still in control of themselves. By stepping in to protect, you most often step on toes. Where does your responsibility to your friends end or begin. By exalting yourself to the point of judge over your peers you must destroy your ties with that person. After taking a step like this the person will no longer be your friend or your parent. Turning someone in is not the same as helping them. Helping them is something that a friend would do; protecting someone is something that a higher being performs. It is the role of government to protect. It is the role of friends to help. Turning someone over to the police is lazy. That is a way to say that you can no longer deal with the problem. If you don't have the guts to stick with someone through a rehabilitation program, then you are not that person's friend. The police are the wrong answer. Once you turn someone over, the matter is out of your hands. If you really care for a person, then it is essential that you help them; not protect them.

Jon Neulieb

Comments

Jon finds himself struggling to figure out exactly why he objects to the idea of turning in drug abusers. The freewriting gives him a chance to develop a line of thought, to find some examples to work from.

ASSIGNMENT

Freewrite for ten minutes on this question: if you discovered a friend using or dealing an illegal substance, would you turn in that person to the police? Why or why not?

Freewriting is an important method not just for writing papers but for helping you to discover what you think about events around you and what your values, ideas, and problems really are.

Freewriting also is an excellent class-preparation method. Any time you must read something for class, a short period of freewriting in response to your reading will help you

- restate the ideas in your own words
- identify points you don't understand clearly
- develop a position on issues.

Writing in response to reading assignments is an excellent study method, better than underlining with a marker.

The following is an example of a response to a demanding reading assignment in a literature class. In it, Jon is clearly preparing himself to take a strong position against Johnson's ideas in class discussion or in an essay. The freewriting helps him to marshal his arguments.

SECOND EXAMPLE OF FREEWRITING

In order to define some new concepts, language must be able to stretch and recombine to give meaning to the new term. The meanings that are culturally inherent in a subterm give the new term subtle nuances and shades that come, with society's use and approval, to mean. That is why you can only guess what Psycholinguistic Genealogy is. This is not an invented term, but, several common terms combined to imperfectly describe a new thought. What the term is going to come to mean as this paper gives it explanation is figuring out who is killing off whose Oedipal literary father. The example that I came across is that T.S. Eliot kills off Milton and then praises the Metaphysical Poets. Samuel Johnson, in his essay "The Metaphysical Poets" kills off the metaphysical poets. In terms of psychology then, Johnson has killed off the metaphysical poets as his father but they were a mother to Eliot.
It is of wonder to me how Johnson can feel his manhood is earned by killing off a woman.
The point is, that those who go before automatically destroy the poetry. Criticism itself may be found to be guilty of this crime. This is why it becomes feasible to critique a critique.
Perhaps it is the position of hindsight, but it seems odd that Johnson's critique would be so essential to how people viewed the metaphysical poets. In short, his critique simply is not that good. Johnson is not only contradictory and selfrighteous but he fails to complete any real argument in the course of his essay.
Some of the language that Johnson uses produces a meaning that is not easily noticible but is clearly destructive. The first glaring problem that arrives is John's use of the term "race".

"... appeared a race of writers..."

This term is used by Johnson as a means of separating himself from these writers. The word is used as a derogatory term to describe the metaphysical poets and brings into question the appropriateness of the term. This kind of usage is not even thought of as peculiar by Johnson and shows his disdain for the even more separated races. I doubt that Johnson would resort to calling the poets any ethnic names, but that is what the usage implies.

The next unclear usage by Johnson is the term "wit." He redefines the term in order to take the metaphysical poets out of the category of wit. In accordance with Pope's definition, Johnson finds the metaphysical poets are witty because "Their thoughts are often new, but seldom natural." Johnson redefines the term in a special method that excludes the poets for obscure reasons. This type of redefining a term is dangerous because it is used to exclude a group that the author finds inferior. Imagine if the government redefined the term "citizen" to exclude all those of minority status. This is an extreme case but it does show the danger inherent in exclusionary or divisory redefinitions of language. Even if students are encouraged to look past Johnson's criticism to observe his prose, his constructions and conceits are much too careless to be admired.

Jon Neulieb

Comments

Obviously, Jon is using his freewriting to blow off some steam against Johnson. While he is doing a good job of gathering his arguments against the essay, he is also moving from an initial emotional reaction ("I don't like this!") to a more rational response ("Here's why I object to this.").

One of the functions of freewriting is to offer an opportunity to get the emotional part of writing out in the open. You can be angry about having to write or angry about some aspect of your subject. Or you can express delight, sorrow, or any other emotion. Freewriting can help you define the emotion and gauge its strength. Sometimes, as in the preced-

ing example, it can help you do the initial exploding before
you have to write more coolly. Letters of complaint are a
frequent occasion for just this kind of writing.

ASSIGNMENT

*Use freewriting to respond to an instance when you have been
treated unjustly, unfairly, or impolitely. Being cheated by someone
you trusted, getting an incorrect bill for the third time, or being
treated as a number by a bureaucrat might be examples. Write
as emotionally as you like. Call the bad guys all the names you
wish.*

*Then go back and write a letter of complaint to the appropriate
person; the letter should be as rational and cool as possible.*

ASSIGNMENT

*Use freewriting to respond to the reading in a class you find
demanding. Develop questions to ask in class, based on the freewriting you do. Freewrite for your class daily for a week. What
happened?*

Freewriting is an important form of invention. Sometimes,
you may need to use freewriting before you use any other
system, just so you can warm up or get past writer's block.
Freewriting is also a good way to remind yourself that you
have to invent, draft, struggle, and wander around your ideas
a bit before you can produce a working version of a paper.
Like other forms of invention, freewriting assumes revision
as a part of the writing process.

Chapter 4

Is there another combination of methods I can try?

Probably you are already familiar with one kind of informal invention, brainstorming. Informal invention focuses on the subject, but it isn't guided by a system of specific questions.

Informal invention is especially useful for getting your mind on a subject, for getting ideas to flow. When you use informal invention like looping or brainstorming, try not to be judgmental. Until you have done the initial invention, make yourself believe that there is no such thing as a stupid idea. In the invention stage, all ideas are equal, so write them all down. You also need to persevere. Keep at it for at least fifteen or twenty minutes; sometimes it takes a little messing around before anything starts to happen, but if you don't write things down, nothing at all will happen.

HOW TO BRAINSTORM

Brainstorming can be an effective method for ransacking your mind for information, whether you do it with a small group of people or alone. While most invention is a solitary activity, brainstorming works quite well when you do it in a small group of three or four people. In that instance, you gain the advantage of others' points of view, ideas, memories, and associations. If you do brainstorming in a group, it is a good idea to appoint one person to write down all the ideas.

Focus your brainstorming on the subject at hand. Don't let the free association technique tempt you to talk about other subjects.

Say your ideas aloud; don't just think them. Articulating an idea automatically sharpens it. Then, once you say the idea, write it down. The idea has had two opportunities to gain shape and strength. Ideas in your head can be exasperatingly elusive, slipping away before you want them to, as other ideas crowd them out. An idea that has been stated, on the other hand, and especially an idea that has been written down, in any shape, is one that can be retrieved, polished, and used.

Brainstorming is especially effective for developing a variety of solutions for complex problems. You can also use it when you have a number of possible subjects for a paper and you want to select one. Especially, you should try brainstorming in a group when you think you have no fresh ideas about a subject.

EXAMPLE OF BRAINSTORMING

In the following example of brainstorming, the students have been asked to think of technological developments that have occurred within the last forty years. The developments may be major ones or relatively minor. Later, the students will be interviewing older people about the changes in their lives caused by one of the developments listed. Note that the students begin with a pair of categories, just to simplify their list making. Then they create the lists. Additions may occur in either list as the session continues. The students may also wish to add a new category at some point. They also add notations about entries that may need reclassifying or further information. Ideas that are immediately related are listed beside each other.

Major developments

air conditioning
jet air travel

computers
stereophonic sound systems (radio, movies, recordings)
microchips (separate from computers?)
television — color tv
fiber optics
space travel & exploration
organ transplants
lasers
polio vaccine (elimination of smallpox?)
antibiotics
communications satellites
nuclear energy
recycling

Minor or related developments

pocket calculators
personal computers
dishwashers
radar detectors
disposable diapers
plastic bags (major? — trash bags, sandwich bags, etc.) plastic wrap
compact disc players
cassette tapes
portable cassette players
vcr's (major?)
garbage disposals   trash compactors
direct dialing and area codes
rack and pinion steering (?)
auto seat belts and shoulder harnesses
microwave ovens

Comments

The brainstorming session has generated a large variety of possible topics for an interview. The next step is to use brainstorming again to select one topic. In this step, the brainstorming will be done individually, rather than in a group. Again, list-making is a key activity in brainstorming, with the addition of notations that clarify something about the list entry for the writer.

POSSIBLE TOPICS FOR INTERVIEW

cassette tapes (especially for cars)
computers. (Uncle Allan's company)
disposable diapers (Grandma and Mom?)
polio vaccine
air-conditioning
tv (Mrs. Jennings?)

The writer can also rank the ideas in order of their appeal. A short note about the reasoning behind the rankings helps the writer later in remembering the direction of thinking on each idea.

1. *tv (Mrs. J. tells good stories about old times)*
2. *disposable diapers (Mom and Grandma easy to reach)*
3. *computers (Uncle Allan easy to reach. Boring topic)*
4. *cassette tapes (our whole house is full) (who would I interview?)*
5. *polio vaccine (who told me that story about an epidemic?)*
6. *air-conditioning (who to interview?)*

ASSIGNMENT

In class, generate a list of technological developments occurring within the last forty years. On your own, brainstorm about persons you know who clearly remember life before some of these developments. Match the two lists as in the preceding example, and then rank the possible interviews according to your interest in them. Make an appointment with your choice of persons to conduct the interview.

You will need to prepare for the interview by developing at least ten or twelve questions. Be sure to take pencil and paper with you so you can take notes easily.

Before you conduct your interview, make a copy of your questions and write what you think the answers will be.

When you conduct your interview, be alert for information you can use besides that given in response to your questions. Watch the facial expressions, gestures, and body language of your interviewee.

Write a paper in which you explain what technological development you have chosen to explore, whom you chose to interview
and why, what you expected, what you discovered, and how the
interview changed or did not change your perception of how technology affects our lives. Do not simply write a verbatim transcript
of the interview; your paper will be a combination of written
dialogue, narrative, and analysis.

How can I use invention to respond to an assignment?

In order to select a helpful invention technique for this assignment, you need to look at what the assignment asks you to
do. One of the steps you have already done, brainstorming
about possible technological developments and people to interview about them. There are two more steps: writing up
the questions for your interview and writing a paper that pulls
all the pieces together. The questions you ask are crucial to
providing you with enough information to create that final
paper. For that reason, you should develop questions that
will give you what the last part of the assignment requires.

Look at the last part of the assignment. Several key words
tell you what kind of information you will need: **what   who
why   how**. Those words are part of a system you probably
already know, the Journalist's Questions.

THE JOURNALIST'S QUESTIONS

Almost everyone knows this invention system. Reporters learn
to answer its questions as they write up a story. The questions
are simple ones, and the set is easy to memorize. Because
it is so portable, the set is a good one to use in situations
where using longer or more complex systems would be diffi-
cult. Essay examinations are one good place to apply this set
of questions; when you are asked to "discuss" something,
use the journalist's formula to make sure your discussion is
comprehensive.

The Questions

Who?   All persons involved in the subject should be listed along with their titles, status, relationships to each other and to events.

What?   The actions involved should be listed, as well as institutions, organizations, and results of actions.

Where?   The place or places involved should be listed.

When?   Dates, times, and lengths of time should be listed.

Why?   Causes and motives should be listed.

How?   Methods, techniques, paths of development should be listed.

These six questions can form the basis for the interview questions the interview assignment requires.

Sample Questions

1. **When** did you first see television? **How** old were you?
2. **When** did you first buy a tv?
3. **Where** were you?
4. **Why** did you buy one then?
5. **What** was your reaction to what you saw?
6. **What** was your favorite program?
7. **What** did you do for home entertainment before television?
8. **How** has television affected your daily life?
9. **How** do you think television has affected larger events, such as politics or international affairs?
10. **How** has tv changed since you first started watching it?
11. **What** problems do you think tv has caused? solved?

Notice that these questions require specific, informational answers. They cannot be answered "yes" or "no," so that the interviewer gets maximum results. Yes-or-no questions often leave you needing to ask another question to get anywhere. The Journalist's Questions, on the other hand, automatically prompt the interviewee to provide solid information. These questions also will generate the kind of information the assignment calls for. The **who** question will be answered in the section of the assignment that requires the name of the person being interviewed and some information about why the writer chose that person. The **when** questions put a date on the technological development, in this case, television. The rest of the questions address the issue of **how** technology has affected our lives. By using these questions, the writer can be sure of meeting the requirements of the assignment.

The following is an example of a student's response to this assignment.

EXAMPLE

JACOBSEN'S THEORY OF TELEVISION

I interviewed Helen Jacobsen at her apartment in Mattoon, Illinois on February 28, 1987. Helen is an eighty-three-year-old lady who can wear me out on an all-day shopping trip. I chose Helen to interview about television because she doesn't watch much television, and I wanted to find out why she listens to the radio more.

Helen remembers life as a sheep rancher's wife in northern California as long hours and hard work. Helen said she really enjoyed coming in from a hard day's work and, after supper dishes were finished, settling down with her husband, Bill, to listen to the radio shows. Bill and Helen would listen while Helen worked on her embroidery and Bill smoked his pipe. Helen enjoyed using her imagination to picture what was happening in her favorite radio shows. Among her favorite radio shows were One Man's Family and the Jack Benny Show, which were comedy shows. Life for Helen was very peaceful and full of enjoyment.

Bill and Helen rarely got to town in the 1940's, because they lived so far out in the country and the roads were of poor quality. They only went to town about once a month for groceries and errands.

On one of these monthly outings Helen remembers seeing television for the first time. While Helen and Bill were getting their monthly groceries they decided to visit friends. This visit was different. The four of them sat in front of a screen and watched other people instead of sitting on the porch and visiting with each other. Helen was disappointed that evening; because she and Bill were so isolated on the farm, she enjoyed talking with other people. Helen said, "This television invention was going to take away one of my most prized enjoyments, visiting with other people." Helen remembers how she and Bill discussed on the way home that evening that they had hardly said twenty words to their friends. Helen said, "I went away as lonely as when I came."

Helen never cared whether she and Bill ever purchased a television. Bill did not share Helen's feelings and purchased a television the next year. Helen doesn't enjoy television as much as she enjoys radio. She said, "I don't care whether I see television, because on radio I picture things and use my mind. I can make all the men handsome. Men's voices can be beautiful, but physically he can be ugly." Helen was very disappointed the first time she saw her favorite radio stars on television: they were not what she had imagined. She enjoys using her imagination while she listens to the radio. The television has taken that away from her.

Helen's husband, Bill, is now dead, and she doesn't watch much television anymore. Helen states, "Bill was the television watcher, not me." Helen sees television as a bad influence over society. Children learn much too much from TV today, too early. Helen said, "Parents have turned their families over to television." It bothers Helen that the immorality on television makes it seem moral to society. Helen said, "The more vile they get, the more people like it."

I understand why Helen doesn't watch much television now. Her theory is that children see things on television before their minds are ready for them. They learn things at much too early an age. She said, "Sex and violence are not for a six-year-old but for an adult. Children are seeing things meant for adults. On the radio," Helen said, "children do not get sex and violence thrown at them. On the radio, children have to imagine it." What I think Helen means is that by listening to the radio young children do not see sex and violence that was intended for adult viewing.

Helen also sees society today as no longer visiting. She said, "Everyone today sits in front of the television and doesn't get out and visit friends and relatives."

Helen sees television as the object that takes people away from her. Helen dislikes television because it makes her lonely.

— Diane Beedy

ASSIGNMENT

As you do the following assignment, you can use the same combination of invention techniques as for the interview assignment.

In a brainstorming session, develop a list of those nagging
little questions you would like answered: Who invented Kleenex?
Why are traffic lights red, yellow, and green? How does the cable
TV company bring in all those channels? Why do the British
drive on the left side of the road? When a dog wags its tail, does
that really mean the dog is happy? How do we know? You can
make up your own list of these questions.

Next, select the question you are most interested in answering.
Either alone or in a small group, brainstorm on sources for the
answer to your question. These may include library sources, but
less orthodox methods are also acceptable.

Develop a list of questions you will need to answer in order
to provide a complete response to your main question. Use the
Journalist's Questions to make up your list.

For example, if you are interested in why the British drive on
the left, you might develop questions such as

> When did the British start driving?
> Who regulates traffic laws in Britain?
> How has the system changed, if at all?
> Where else besides Britain do people drive on the left?
> Why don't the British change over to the right-hand system?
> What caused the system to develop in the first place?
> What advantages are there, if any, to driving on the left?

Having specific questions to ask will make your research easier.
It will help reference librarians as they guide you to appropriate
sources, and it will suggest key words to use in indexes and
computer searches.

Write a paper in which you provide a clear, factual, and
documented answer to the question.

The combination of brainstorming and the Journalist's
Questions works especially for developing term paper topics
and then for focusing your research. Often students faced
with term paper assignments struggle with the enormous volume of material they find in college libraries. If, however,
you are looking for the who, what, when, where, how, and

why of a subject, you can recognize useful information when you find it, rather than wondering if that article on the basket-making techniques of South American tribes is really what you need.

Before you head for the library, make up a list of the specific questions you want to answer. Mark off each question as you find the answer, so that you can track your progress through the project. Be sure to write down the exact source of each answer.

Is there another method I can try?

Yes. More informal invention combined again with the Journalist's Questions can help you break up mental logjams.

HOW TO CLUSTER IDEAS

To create clusters of ideas, write down a word or phrase in the center of a piece of paper (or in the middle of the chalkboard in your classroom). The word or phrase should indicate the general subject you intend to write about. Around this central idea, write more words and phrases indicating parts of the subject. Continue to write down all associated ideas or words that connect either to the main idea or to a smaller, more specific part of the idea.

The group of ideas that seems the most specific or that generates the most items may be the one you should examine more closely.

Clustering is an excellent technique for finding a subject for a paper.

EXAMPLE OF CLUSTERING

Athletes	Teachers	Presidents
Walter Payton	Mrs. Jones	Lincoln
abilities	knew history	freed slaves

Athletes	Teachers	Presidents
running	competitive	Gettysburg
pass receptions	fair	Roosevelt
reputation	"Rocky"	Lend-Lease
records	patient	Depression
charities	fair	WWII
		NATO

This example looks like a series of outlines, but that is only one way to cluster. You should do it on a big, blank area, like a sheet of paper or a chalkboard. You can use circles and arrows to connect ideas to each other. The point of clustering is to get ideas down so you can look at them, even in the rough form they'll probably take. Use words or phrases or even symbols. Clustering is not a place for worrying about complete sentences or correct grammar and spelling.

The example shows how one student approached the problem of identifying people he admires. First, he thought of several categories of admirable people. Then he tried to come up with specific persons under each of those categories. Then he listed some of the things that made him admire each of those people. The items are indicated by short words or phrases that are meaningful to the student. NATO, for instance, suggests the whole matter of Roosevelt and Churchill establishing the basis for the North Atlantic Treaty Organization. But in clustering, a key word, phrase, or symbol is enough. All you need is a trigger for your ideas.

The cluster includes information that comes to mind quickly and easily. One purpose of clustering is to gather the facts that you know as fast as you can and to lay them out where you can look at them and at their relationships.

Try to identify how your ideas relate to each other: cause and effect? categories and subcategories? comparison? contrast? sequence?

Another look at the cluster indicates several paper possibilities. For example, the student could write about either of his former teachers and the importance of fairness as a quality in a teacher. However, this student decided to write about a person for whom he felt strong admiration, Walter Payton of the Chicago Bears. The fact that he listed more about Payton than about any of the other possibilities suggests that this was a wise choice.

EXAMPLE

SWEETNESS

There are many athletes whom I admire. The one who truly stands out, both on and off the field, would be the man Chicago fans call "Sweetness." Yes, I'm talking about the great Walter Payton, a twelve-year running back of the World Champion Chicago Bears. Payton owns many Bear records as well as NFL records, such as being the league's leading career rusher. I can still remember coming home from church at the age of eight and watching my favorite player in action. His enthusiasm and his awesome running ability mark him as one of the best running backs ever to play the game.

His style of running isn't glamorous, but it's so outstanding that even opponents appreciate him. Even in high school, Payton's attitude and skillful running had a powerful effect on those around him. I like Payton's style of play because when linebackers hit him as hard as they can, Walter pops right up for the next play. Payton hasn't missed a game due to injury for over 12 years.

Walter Payton is truly an outstanding player. In his 12-year career, he has broken over 20 NFL records and 10 Bear records. John Madden, an ex-coach of the Oakland Raiders and present TV announcer, calls Walter Payton "the most complete player ever to play in the NFL. This guy can do it all," Madden explains. At the rate he's going, he may pass the 20,000 yard mark before he ends his career, unless he gets hurt.

Walter Payton is not only an outstanding player, but he is a true gentleman. He took time from his busy schedule to film two United Way commercials without pay. I think that it is important for a professional athlete to take the time to help others. In all the years that I have watched Walter Payton play, I never saw him lose his temper or argue about a call *by an official. He is a great example of how the game should be played.*

I just love to watch Walter Payton run with the football in one hand as if the ball is a piece of fruit. I think that the name "sweetness" describes Walter Payton as an all-around person not only in sports but in life.

There are not many players like Walter Payton. I don't see how any person can say that Walter Payton is not the best running back in the history of the NFL. He's an outstanding runner with a good sense of humor. For instance, when he broke the rushing record, he received a phone call from Ronald Reagan. Without batting an eyelash, Payton said into the receiver, "The check's in the mail." Then Payton was asked who he thought would break his record. He said that he didn't care who broke his record "as long as it's my son." There is no doubt in my mind as well as in many others that Walter Payton has a spot in the NFL Hall of Fame.

—*Mark Roedder*

How do I revise from clustering?

The Journalist's Questions will work very well for revising a draft such as Mark's. Mark has begun analyzing the admirable qualities of Payton, but it is clear that he needs to develop his subject more thoroughly. The problem is one of specifics, and the Journalist's Questions can help him gather those details. For example, Mark can ask, "**What** specific records has Payton set?" "**Where** did he set them?" "**When** did he set them?"

ASSIGNMENT

Use clustering to identify qualities and persons you admire.

Use the Journalist's Questions to describe someone you admire. The person should be living and should not be a close relative. Using the information you gather in response to the questions, make clear to your reader why you admire this person.

EXAMPLE

LEARNING TO BE AN RA

[Who?]   *Angie McGowan is my RA for Taylor Hall north, on the*
[Where?]   *second floor. She has been an RA for one year and this is*
[When?]   *her second. Angie really likes being an RA because of the*
[Why?]   *friends she meets. She likes to help people because it gives*
           her satisfaction.

           Getting started was very easy. All Angie had to do was
[How?]   *go through orientation during the summer for a week. She*
           had to put in an application and go before a board and get
           an interview.

[Why?]   *Angie says the pay is good. She gets free room and board*
           which is very helpful since school is so expensive. She also
           gets $160 a semester, or $80 every two months, which takes
           care of her spending money.

[What?]   *Being an RA is very time consuming. The kinds of things*
           Angie has to do are programming, counseling, duty, and
           administrative work. Some other things she does are
           personal projects. For example, making signs for people
           when it is our birthdays and decorating doors, making
           friendly visits, and giving out Halloween candy.

[What?]   *The one greatest thing Angie has gained from her job is the*
           friends she has made. Her job has many advantages, like
           teaching her responsibility, time management, listening
[Why?]   *skills, organization and dealing with all kinds of people.*
           I could tell that she had many more ideals but she was
           rushed for time.

           Some disadvantages of her job are enemies, having a hard
           time working with peers, and sometimes getting too
[What?]   *emotionally involved.*

[Why?]   *One terrible story Angie told me about was one time a girl*
[Who?]   *on the floor tried to commit suicide. Angie said another*
[Where?]   *girl on the floor told her that her roommate had taken 142*
           aspirin. Angie said she stopped being an RA for the
           moment and was a friend. She talked to the girl and took
           her to the hospital. The girl is fine now and from time to
           time Angie goes to see how she is doing.

[How?] *I can tell Angie likes her job very much, since she always*
        has a smile on her face. She lives right across the hall from
        me and comes to see me often. She is more to me than an
        RA, she is a true friend.

                                                           — Diane Richardson

Comments

Diane identified a person a little closer to her life than Walter
Payton to write about. Her clustering used the same principles
as Mark's, however.

Diane has used the Journalist's Questions to gather information about her Residence Assistant. She succeeds at providing
the basics; we know who the RA is, where she lives, and how
Diane knows her.

Diane can use the same questions to expand the paragraph
on the problems an RA faces. While the next paragraph, the
one about the suicide attempt, tells something about those problems, Diane mentions others that we hear nothing more about,
such as the problem of making enemies. Diane could ask herself
some more questions:

Who are these enemies?
Where does the RA encounter them?
Why are they enemies?
What things do they do?
How do they act?

We see how the RA coped with a suicide attempt. If we can
also learn about her enemies and how she copes with them, we
will, perhaps, understand more about why Diane admires her RA.

USING THE JOURNALIST'S QUESTIONS TO REVISE
SENTENCES FOR DETAIL

The Journalist's Questions also work when you need to expand
on your ideas at the sentence level. If your teacher writes
comments such as "more concrete detail" or "be specific" or
"tell me more about this" in the margins of your papers, use
the Journalist's Questions to take care of the problem.

The following is how to revise your sentences for concrete
details.

Begin with a simple sentence or base clause.

She hit him.

This sentence doesn't tell us much except that a female of some kind has hit a male of some kind. Add a word that tells **how** she hit him.

Abruptly she hit him.

or

She hit him abruptly.

Now tell **what** she hit him with.

Abruptly, she hit him with the news.

Now tell **where** she hit him.

Abruptly, she hit him with the news at the breakfast table.

Now tell **why** she hit him.

Abruptly, she hit him with the news at the breakfast table, hoping that he would actually listen at that hour.

You can see how much more information has been added to the sentence by using just four of the seven possible questions.

TRY TO ANSWER AT LEAST ONE OF THE JOURNALIST'S QUESTIONS IN EVERY SENTENCE

While you need to vary sentence length and while not all sentences need to be long and full of information, your sentences do need to keep your paper moving. One way to make sure that happens is to use the questions to add more information as you write each sentence.

ASSIGNMENT

Revise your last essay by answering at least one of the Journalist's Questions in each sentence. Thus you should add at least one item of new information to every sentence you wrote. Write the question you answered in the margin by each sentence. Which questions did you ask most frequently? Which did you ask least frequently? Notice your pattern; you may need to practice asking that least-frequent question more often.

Chapter 5

Once you have become comfortable with using invention methods, you can extend your repertory a little. There are many methods to choose from; you will probably find two or three you prefer. Even so, it is handy to be familiar with others, for those occasions when a particular system seems most appropriate for your needs. A good invention technique is one that gets you the results you need.

How else can I use invention?

Writing tasks for college students often require analysis and evaluation. Examples include book reviews, analytical papers about literary or musical works, evaluations of journal articles, reviews of professional literature on a particular subject. Students often struggle with these tasks because they don't really know how to proceed. Teachers sometimes provide models for what the final paper should look like, but in classes where writing is not the main emphasis, you may wonder how to conduct your own analysis or evaluations before you produce the final draft for class. Two invention systems, one new and one very old, can help you solve this problem.

AN OLD SYSTEM — THE TOPICS

The old system dates back to Aristotle. He developed a system called *topoi* for finding arguments for debates and discussions. The word *topoi* means the places to look. We now call these "places" topics. In essence, the topics are a series of mental activities that you record in writing. The thinking you do as you work through the topics forces you to analyze your subject in exhaustive (some say exhausting) detail. The topics are exacting and sometimes tedious, but they yield first-rate analysis. When you really have to demonstrate that you know your subject, the topics are the method of choice.

While the topics are a formal system like cubing or the Journalist's Questions, you can be flexible about how you use it. When you are struggling to understand a subject, apply the discipline of the topics to the subject to help you get a grip on things.

System D works fine with the topics — whatever you write will be helpful in some way. There are just two kinds of invention: good and better.

Often you won't have to use the whole system of the topics. Writing teachers for years have used an approach called "the modes," which simply means giving writing assignments that emphasize one of the topics, such as definition or comparison/contrast. If you encounter one of these assignments, you only need to use that part of the topics that is appropriate.

RECOGNIZING THE MODES

Look for assignments with phrases or words like these:

Compare and contrast
Classify or write a classification of
Define or write a definition of
Analyze or write an analysis of
Discuss the causes of
Discuss the effects of
How does _____ affect _____?

Many essay questions on examinations also require you to use only part of the topics, questions that ask you to classify, define, compare, or discuss cause and effect. As you can see, much college writing draws directly on the methods of the common topics, so it is certainly worth your while to be familiar with how they work.

Recognizing the Topics in Essay Questions

Look for essay questions with phrases or words like these:

Compare _____ with _____.
Contrast _____ with _____.
Show the difference between _____ and _____.
Discuss how _____. [Cause & Effect: Antecedent &
    Consequence]
Define _____.
Classify _____.

When you encounter these essay questions, plan to use the system described in the following section. Teachers who ask these essay questions generally expect you to provide definition, classification, and so on in the way described by Aristotle.

The systems you have used up to now are easy to memorize and carry around in your head; they are "portable." The common topics are less easy to manage because there is considerable detail under each topic. The best way to handle this problem is to master one topic at a time.

THE COMMON TOPICS

CLASSIFICATION — selecting a feature that all the items in a group have and then examining the variations of that single feature

1. Develop a principle of classification, a characteristic which you consider in all the items being sorted.

    EXAMPLE
    Cars may be classified by their sizes — subcompact, compact, mid-size, full-size

2. Stick with your principle. Changing principle of classification will cause overlap among your categories.

    EXAMPLE
    You cannot categorize subcompact cars, compact cars, and red cars because some subcompact cars are also red.

3. Use a principle that will account for all the items being classified. If you come up with a category labeled "miscellaneous," your principle may need review.
4. Avoid good, bad, and indifferent as categories. These labels are matters of opinion instead of observable fact. Also, they tend to disguise a real principle of classification.

EXAMPLE
Calling a movie good may mean that it is a vampire movie or it may mean that it is beautifully photographed. You need to know what principle is being applied before you know whether to believe the statement that the movie is a good one.

DEFINITION — explaining the terminology you are using, especially terms your audience may not know or terms you are using in a specialized way.

Definition can be as straightforward as a one-sentence, formal definition or as complex as an extended definition. The tougher the concept or term you are using, the larger the definition your audience will need.

1. *Formal definition* — one sentence including the term to be defined, the category to which the term belongs, and the differentiating characteristics that distinguish the term from other members of its category.

EXAMPLES
Grading [term] *is a means of indicating the quality of students' work* [category] *by dividing the levels of quality and assigning each a numerical or letter value* [differentiating characteristics].

An alligator [term] *is a large, carnivorous, four-legged* [differentiating characteristics] *reptile* [category] *native to the semitropical regions of the southeastern United States* [differentiating characteristics].

2. *Extended definition* — a longer definition containing a

variety of explanations for the term. Especially appropriate for abstract terms.

a. Provide a formal definition.
b. List some synonyms.
c. Give the etymology or origin of the term.
d. Compare the term to something else.
e. Contrast the term with something else.
f. Give an example.
g. Tell what the term is **not** (negation).

EXAMPLE

Hazing [term] *is an initiation activity* [category] *in which students engage in horseplay, practical jokes, or humiliating ordeals in order to be admitted to a group* [differentiating characteristics]. *Hazing is a way of hassling or teasing students to test their commitment to an organization* [synonyms]. *The word* hazing *comes from the Old French* haser *meaning to irritate* [etymology]. *Hazing is like the "I dare you" games young children play to test each other's social acceptability* [comparison]. *Hazing is not like the pleasant experiences that build friendships, such as going fishing together or building something* [contrast]. *A dangerous form of hazing practiced among some fraternities involves requiring pledges to drink huge quantities of alcohol at a single sitting* [example]. *Hazing is not an officially accepted form of admitting people to membership in fraternities and sororities* [negation].

COMPARISON AND CONTRAST — comparison requires you to consider how two things are alike both literally and figuratively. Contrast is considering how they differ or resemble each other both in kind and in degree.

How are they alike?
How are they different?
How much are they alike?
How much are they different?

EXAMPLE

Ballet and square dancing both use sequences of named steps, such as arabesque and dosido [literal comparison]. *Ballet resembles French cuisine because it is formal, traditional, and carefully structured. Square dancing contains some of the* same elements of frontier life as American cooking; it is spontaneous, energetic, and created from the elements at hand [figurative comparison]. Both ballet and square dancing make use of male-female partnerships as well as ensembles [similarities in kind]. In ballet, there are clearly identifiable star dancers who perform solo during the dance. In square dancing, no one dances solo [differences in kind]. Both ballet and square dancing require mastery of specific steps and development of strong partnerships, their most important similarity [similarity in degree]. Ballet takes years of training and discipline and only a few dancers ever achieve distinction, but square dancing is available to people of all ages and physical strengths [difference in degree].

CAUSE AND EFFECT — discovering why and how things happen and tracing or predicting results

Cause   *What caused _____ to happen?*

What evidence supports the belief that this cause is the correct one?
Is there any evidence to the contrary?
Were there multiple causes or a single cause?
If there were several causes, can they be ranked according to their importance?
If the causes are not definite, make a list of possible causes.
Are any causes more probable than others? Why?

Effect   *What is the effect of _____?*

Is there only one effect, or are there several?
What evidence indicates that these are the effects?
Is this effect possible?
Is this effect probable?
Does the effect logically follow the cause? [Remember that simple sequence does not necessarily imply a cause–effect relationship.]
Can the effects be ranked according to their importance?

EXAMPLE
What caused the drought of 1988 to afflict a large area of the central United States? Some scientists argue that carbon emissions from automobiles, factories, generators, and so on

have filled the atmosphere, creating the "greenhouse effect" of
trapping heat in the earth's atmosphere. Evidence to support
this theory includes a gradual warming trend worldwide over
the last decade or so. The effect of humans on the environment is becoming more apparent; waste disposal problems,
destruction of wildlife habitat, toxic chemicals, and nuclear
accidents all suggest that human technology can have large-
scale, dangerous effects.

On the other hand, scientists studying the possible effects
of a nuclear holocaust have predicted so-called "nuclear
winter," a cooling off of the earth because of enormous
amounts of material in the atmosphere blocking off the sun's
light.

The drought could also be caused by the "El Niño"
phenomenon in the Pacific Ocean. The changes in atmos-
pheric pressure and temperature of ocean currents caused by
El Niño affect climates all over the world, and in 1988
drought conditions existed in many places.

The effects of the drought will certainly include higher
food prices. Other effects include strains on systems that
deliver electric power; the demand for air conditioning is
very high.

On the other hand, one good effect may be greater water
consciousness. Americans waste water; the drought may help
change our patterns of using water. An example is the way
we maintain lawns. Even though lawns are an English
invention, people in all kinds of un-English climates want
smooth, green lawns. But does it make sense to pour gallons
and gallons of water on a lawn in Phoenix to keep it looking
like a lawn in Kent? If we are running out of water, aren't
lawns an expensive indulgence?

COMMENTS

While this writer is starting to change the direction of his
invention at the last, that is not necessarily bad. **Remember
System D.** Invention exercises often have the effect of leading a
writer into surprising territory. Certainly the cause–effect material
at the beginning is fairly specific, but it is also predictable —
straight off the evening news. When the writer starts discussing
the drought in terms of people's lawns, however, he provides an
interesting and fresh approach to the problem.

ANTECEDENT–CONSEQUENCE — exploring what might follow from a given set of circumstances; often cast as an "if-then" statement.

EXAMPLE
If a parent agrees to a midnight curfew for a child upon her sixteenth birthday, then the parent must grant the same curfew to younger children when they each turn sixteen.

The logic of this statement breaks down into these steps:

The oldest child receives permission for a midnight curfew upon turning sixteen.
All the children in the family should be treated equally.
Therefore the younger children can expect the same curfew when they become sixteen.

EXAMPLE OF ANTECEDENT–CONSEQUENCE:
If we stop putting expensive green lawns around our houses, what would be the consequences? Something would have to replace the lawns; lawns are a form of territory announcing whose place this is. We would use less water, but green grass cools the air and cleans it. Grass produces oxygen, holds the soil, and doesn't absorb heat like asphalt. So the stuff we replace lawns with will need to perform those functions. If we save lots of water, will we still come out ahead?

COMMENTS
The writer is still struggling with the lawn problem and has started asking some important questions. Notice how the topics of cause/effect and antecedent/consequence apply to problem-solving. In this instance, the writer has not come up with clear answers, but the questions are significant and make for interesting reading, which is the goal of invention in the first place. What he **has** done is to develop a subject for a unique term paper, exploring alternatives to grass lawns for home landscaping. The common topics, as structured as they are, can still be open-ended and flexible. This writer is following his nose toward an interesting approach to his subject, trusting the unexpected. System D and successful invention depend on that kind of trust.

THE COMMON TOPICS IN A NUTSHELL

Classification
Definition
Comparison and Contrast
Cause and Effect
Antecedent and Consequence

These are the most commonly used of the topics. One other is especially useful when you need to develop arguments in support of a position.

AUTHORITY — in ancient times, a person revered by the community for goodness as well as wisdom; now defined more in terms of scholarly credentials or experience. In either case, the reputation of the authority is important.

Has the authority been correct on similar subjects?
Have the authority's methods been tested?
What are the authority's credentials?
Has anyone contradicted the authority? Why? How?
How recent is the authority's opinion?
Has any other view superseded it?
From what assumptions does the authority proceed?
Are the assumptions justified?

College students need more opportunities to use the topic of authority. While you may be introduced to it in an English class, chances are that you can use it frequently in other courses. Most students come to college accustomed to acquiring information — dates, names, places, events, and so on.

You may continue this kind of learning for a year or two, but the further along you get in your college career, the likelier it is that you will be looking at the variations among the experts. You may, for example, be asked to compare theories of cosmology or two historians' views of the Eisenhower presidency or several explanations for a species' extinction. In these instances, you may or may not be told that you need to examine the authorities you consult, but you will be expected to do it regardless. Your college catalogue probably doesn't say so, but your ability to use the topic of authority is a key to academic success.

Given the need to use this topic, you also need to know how to find the answers to the questions it poses. How do you find out about contradictions to an authority's opinion and so on? It would be nice if there were an easy way to do that, but frankly, it involves hard work.

Evidence of controversy tends to turn up in predictable places. For example, books that present revisions of older ideas usually announce that fact in their introductions or forewords, so it is always a good idea to read those parts, even in the textbooks you use. Footnotes or endnotes also provide evidence. Scholarly articles almost always begin with a section called a review of literature, in which authors make clear that they have read all the pertinent material; it is a way of presenting their credentials. If a subject has been controversial, the review of literature should make that apparent.

It is also useful to check the date of publication for a work; the older it is, the likelier that a revision of the ideas has occurred. In fact, you can just about count on any idea being subject to review if its source is more than five years old; a very old source needs to be considered more as an artifact of its time than as representative of current thinking.

Finally, you should check the credentials of the author of an article or book. Usually, an author's association with a college or university will be listed, and the author's academic preparation will be described. If this information is not provided or if the author has a degree listed with no institution mentioned, you should look for other evidence that the author is a knowledgeable source.

WHERE TO CHECK AN AUTHORITY

Introductions	"Review of Literature" Sections
Forewords	Date of Publication
Footnotes and Endnotes	Author's Credentials

You need to determine whether the writer is reporting firsthand information or secondhand information. Reading introductions, forewords, footnotes, and so on does not usually make for an exciting time; some of it may be incredibly boring and tedious stuff, especially those endless "review of literature" sections, but taking the time to check out these sources will reward you.

Keep in mind that scholarly wrangling in print is more common than agreement on the facts. The habit of questioning leads to important discoveries. The topic of authority is as much a matter of asking "What if X is wrong about this?" as it is asking "Who knows the most about this?"

For example, until the sixteenth century, Aristotle was the acknowledged expert on the natural world. If Aristotle said it was so, nobody questioned him. In the sixteenth century, though, somebody did question his authority, and now we know that the world is round and that it isn't the center of the universe.

How can I use the topic of authority?

You can use the authority topic in three ways. First, you can use it in argumentation to strengthen your position. Citing an authority is usually a compelling argument; it becomes more compelling if you can point out that your authority is widely respected and has heavyweight credentials.

Second, undermine your opposition by pointing out weaknesses in the authorities they use.

Third, in expository or informative writing, citing various authorities makes your work more convincing. Your reader (usually an instructor) will pay attention because you clearly have done your homework.

THREE WAYS TO USE THE TOPIC OF AUTHORITY

Mention an expert's view
Refute an expert's view
Become an expert yourself

A more subtle effect of using the authority topic is that it makes you into an independent thinker. You don't have to believe a thing is true just because someone says it is. The habit of weighing the quality of your information is one of the expected results of a good education; this topic is the means to acquire that habit.

ASSIGNMENT

A. Read the introduction to the textbook in one of your other classes. Look for statements that the book's authors are either questioning or supporting an earlier authority.

B. Look in your textbooks for information about the credentials of the authors. Where did you find it? What were the credentials?

C. Make a list of ten magazines. Write a one- or two-sentence description of the point of view each one seems to represent. For example, what is the point of view of Ms. Magazine? Newsweek? TV Guide? People?

D. After you have made your list of magazines, make a list of those you would mention in a term paper. Which ones would you **not** use in a class assignment? Why not?

ASSIGNMENT

Select a movie that has been in theatrical release (not a made-for-TV movie). It may be a new movie or an old one, but you should have seen it recently enough to have a clear memory of it.

Write a statement about the movie in which you do the following:

Classify the movie (comedy? action/adventure? horror? western?)

Explain your criteria for a good movie, and define any specialized terminology (special effects, editing, cuts, film noir, etc.)

Rate this movie according to your criteria.

Read five reviews of this movie. You will need to know the year of the movie's original release in order to find reviews in the

*library. Try to find a range of review sources — popular magazines,
newspapers, family magazines, and perhaps a tabloid. For example, your reviews might come from* Newsweek, Parents' Magazine,
the Chicago Tribune, New Yorker, *and* Rolling Stone.

*Summarize the contents of each of the reviews, citing the
source for each one. [You may need to consult your handbook for
how to document sources such as these.] Note the difference in
approaches; for example, a reviewer in* Parents' Magazine *might
criticize a movie for excessive violence while a reviewer in* Rolling
Stone *might be more interested in the cinematography and consider the violence merely part of the plot.*

*Write a statement in which you discuss the effect of the reviews
on your opinion of the movie you chose. Has your attitude toward
the movie changed? How? Should your readers see the movie or
not? Use the authority of at least one reviewer to support your
recommendation.*

Example

FOOTLOOSE

I really enjoyed the movie Footloose. It was about a guy
who came from a big city life to live in a small country town.
The whole town was centered around the church. Drinking
and music were not allowed because of an accident that had
happened a few years back. It involved a group of teenagers.
The guy who came to live here was used to hearing music,
being able to drink, and just being able to do what he
wanted to do. He decided to try and get the board members
of the town to let the senior class have a prom. All of the
kids were behind him, but the preacher was bound and
determined not to let them have a dance. The guy spoke
before the board, and it ended up that the preacher let the
kids have a senior prom. It had to be in a grain bin on the
other side of their town.

In Christianity Today (April 84, 49–50), Billingsley states
that Footloose is a sort of film that sets off critics but makes
lots of money anyway. Billingsley says, "The only redeeming
performance is by John Lithgow as the minister. In spite of a
truly execrable script he evokes some sympathy for the man;
one can almost believe he is real. The other characters are the

stuff of cartoons." This reviewer thinks that the music in
Footloose is banal and the dancing is sloppy.

Harry M. Cheney agrees with Billingsley. "The characters
in *Footloose* do not resemble kids so much as prisoners of
war. All of the joy has been taken from their lives," thinks
Cheney. "The kids pursue pleasure with grim determination.
Their final victory is simply one of self-absorption over
misguided principles" (*Christianity Today*, April 20, 1984,
49–50).

K. Grubb is also negative. He tells a lot about the movie
itself. "The film loses itself in so many internal contradictions
one is tempted to call it Screwloose." The minister forbids his
daughter to listen to rock, but allows her to do such things as
wear a lot of makeup, dress like a big city hooker, and to do
her hair as a glitzy country songstress. "Even though the kids
have been denied dancing for a long time, they are as slick as
regulars on Soul Train" (*Dance Magazine*, May 84, 149–51).

On the other hand, Stoop explains, "The script of
Footloose was about young lives and had a kind of energy,
restlessness, and honesty that I responded to. I wasn't
pressured by either high budgets or big stars. *Footloose* tells a
story of a hip Chicago teenager whose mother is relocated in
a rural community of Utah that frowns on contemporary
music and bans dancing" (*Dance Magazine*, March 84,
58–60).

Ansen disagrees with the earlier reviewers. He states that
Footloose is about a minister who rails from his pulpit about
obscenity and rock-n-roll and wants to ban "dirty books" from
the school library. "*Footloose* has a lively, sweet, infectious
spirit, and for that one is willing to overlook some clunky
scenes, fuzzy motivations, gracious brawls and the failure to
evoke this town with any sociological coherence. *Footloose* is
notable as a variant of the new Paramount musical style"
(*Newsweek*, Feb. 20, 1984, 78).

I disagree with the reviewer on his thought that the
dancing was sloppy and the music was banal. I thought the
dancing was superb and the music was upbeat. I agree with
the last reviewer because he had a positive attitude toward
the movie. I especially liked the final scene which was at the
prom. There was a lot of dancing and it just really made you
want to get up and start dancing yourself.

My feelings have not changed since I have read the
reviews. I still have the same feelings now that I did before I

read the reviews. I think my opinion has been reinforced somewhat because I know a little more of what the characters are really like. By hearing other people's opinions, I have a clearer view of the movie itself.

I highly recommend that anyone see the movie _Footloose_. It has a good meaning in it. It goes to show that teenagers sometimes have the right idea about how things should be done. _Footloose_ truly is a movie that shows how kids pull together when they feel strongly about such a thing as dancing and music. It describes how a whole town has to move on after a terrible thing happens. The rest of the kids shouldn't be punished and deprived of things they like to do just because of one accident. I think _Footloose_ is a movie the whole family should go see.

—Misti Cox

Comments

This writer uses authority in at least two ways. She cites a variety of reviewers, both positive and negative. Her choice of _Dance Magazine_ as a source for a review suggests that she sees the movie primarily as a vehicle for dancing and music, not for story-telling, though later, she shifts back to an emphasis on story, suggesting that in another draft of the paper she will need to select a focus. She also contradicts authority, showing that authority can be refuted or questioned.

The topic of comparison also appears in this draft. The writer uses direct quotes to show how the various reviewers agree or disagree with each other, so that differences in degree and in kind are both addressed.

Misti also uses the topic of cause and effect. She tells why she agrees with some reviews and not with others. She discusses the effect of the reviews on her as well as the effect of the movie itself. The sections using this topic need more development. A cause-and-effect relationship is described in these instances, but it is not explained or justified very extensively. In a later version of this paper, an important task will be to develop the cause-and-effect topic.

Misti could also use definition and classification here. She sees _Footloose_ as a dance movie, while many reviewers she cites talk only about its improbable story. A definition of dance films could solve her problem of needing to explain more fully why she disagrees with the reviewers. Dance movies often have improbable plots and implausible situations, but the audiences don't care.

Misti could describe at least two categories of dance movies: those in which plot is secondary to the dance sequences, such as *Footloose* or *Dirty Dancing*, and those in which the plot comes first, such as *The Turning Point* or *White Nights*. The assignment calls for the movie to be classified, something Misti does not do in this draft. If she does classify *Footloose* as a dance movie intended primarily to provide a showcase for music and dancers, she will solve several problems with this version of her paper.

ASSIGNMENT

Revise Misti's draft, incorporating some of the preceding suggestions and any others you think of. Discuss your revisions in class.

INVENTION IS A MAJOR AID TO REVISION.

USE INVENTION TO REVISE AS WELL AS TO DRAFT A PAPER.

Twenty Questions for the Writer

In addition to the formal topics, a more modern invention system is an excellent one for developing ideas on any topic. It, too, will work for term papers and other complex writing assignments.

The Twenty Questions will work, also, when you must revise. It is a good system for filling in the gaps that your reader will want filled.

A modern writer, Jacqueline Berke, has developed this system that offers many advantages to student writers.* The "Twenty Questions for the Writer" neatly combine elements of the journalistic formula (who, what, where, when, why, how) with elements of the common topics. The questions are stated simply, so you should be able to use them easily.

Some of the questions in this system do not appear in any

*Berke, Jacqueline. *Twenty Questions for the Writer: A Rhetoric with Readings*. 3rd ed. New York: Harcourt, 1981.

other, so they can help fill in some gaps in your exploration of a subject. For example, one question asks for a summary of your subject. Another asks for your personal response to the subject. Even more important, two of the questions ask you to contrast the difference between expectation and fact— how something should be done compared to how it actually is done.

You may not want to mess with answering twenty questions. Don't be discouraged by the length of the list; some of the questions will require only short answers and others won't apply to your subject. It is important, though, to answer all the questions which **do** apply and to answer them as completely as possible.

USE THE TWENTY QUESTIONS TO

1. Explore a topic for the first time
2. Test the extent of your knowledge of a subject
3. Find a thesis statement
4. Plan research for a term paper or report

We will see how the questions apply to these problems.

In the questions, X stands for the subject you are writing about. Thus, you can substitute a word or phrase for X, such as "junk food" or "my favorite movie." You don't even have to make the phrase graceful or extremely specific. Working through the questions may help you clarify the exact topic of your writing.

THE TWENTY QUESTIONS

What does X mean?

How can X be described?

What are the component parts of X?

How is X made or done?

How should X be made or done?

What is the essential function of X?

What are the causes of X?

What are the consequences of X?

What are the types of X?

How does X compare to Y?

What is the present status of X?

How should X be interpreted?

What are the facts about X?

How did X happen?

What kind of person is X?

What is my personal response to X?

What is my memory of X?

What is the value of X?

How can X be summarized?

What case can be made for or against X?

Example

What does "junk food" mean?
It is a phrase to describe food that has no real nutritional value. Most of it has lots of calories but few vitamins or proteins.

How can "junk food" be described?
JF often has a lot of sugar in it, and it usually is packaged in cellophane wrappers. It may also have a lot of preservatives or even "artificial flavorings." Fillings and frostings are made of strange chemical combinations with things like sodium caseinate. The filling in a Twinkie isn't really cream; it's "creme," which just means it's white and sticky and sweet. Other junk food may be loaded with salt and oils, like peanuts or potato chips. Potato chips, creme-filled cupcakes, candy bars, designer ice cream bars, soft drinks and "fruit-flavored" juices are all examples of junk food.

What are the component parts of junk food?
Salt, sugar, preservatives, artificial flavorings, food dyes, oils, sometimes flour, sometimes fruit or fruit flavorings.

How is junk food made?

In a chemistry lab with bits of kitchen equipment. It is mass produced. I wonder if human hands ever touch it. Advertisements suggest that it is made by elves or grandmothers or little girls dressed in gingham or Scandinavian gourmet cooks. My grandmother specialized in fattening stuff, but she also believed in making me drink my orange juice and eat my spinach. Her fattening treats, like Baltimore cake or southern fried chicken, had some nutritional value and they didn't have preservatives in them. I know grandmothers and elves don't make this stuff. Chemists make it.

How should junk food be made?

Assuming that we can't give it up completely, I guess it should be made with more attention to its nutritional value. Is low calorie junk food really safe? Or are we trading one set of dangerous substances (too much sugar, too much salt, too many preservatives, too much fat) for another set (saccharine, etc.)? At least it should be labeled so that people who don't read lists of ingredients will still have a chance to notice that the stuff isn't especially good for them. Maybe a label on the cellophane that reads "If you eat this it will make you get fat, make your teeth rot, and cause acne."

What is the essential function of junk food?

Junk food doesn't keep us alive or anything like that. It is meant to provide pleasure. We like the taste of it. We may even like the texture of it even though I think those creme fillings are sticky and gooey and nauseating. But I love sour cream- and chive-flavored potato chips, even though I am not really sure they are made from real potatoes and I know no sour cream was in the same lab with them. Junk food is a taste treat. We like the smells sometimes too. I love the smell of chocolate, the taste of it, the texture of it. Chocolate isn't good for me, but I get enormous pleasure from it. Junk food is about fun, not about being good for you.

What are the causes of junk food?

I don't know. Hungry college students and lots of parties, which make a market for junk food. Do they have junk food in India? Is it an American idea? We create a lot of products and then create markets for them, things that aren't necessary and don't really improve anybody's life. Is junk food like that?

What are the consequences of junk food?

A lot of overweight people. People who may eat lots but who are undernourished or who have vitamin deficiencies. People with cavities. Happy people loving their chocolate, creme-filled gooey yummies. People stuffing themselves merrily with designer potato chips or a $3 ice cream confection made from Swiss chocolate, real cream, and almonds. You can't be mad at the world while you're eating one of those.

What are the types of junk food?

Sweet stuff and salty stuff. The sweet stuff includes soft drinks and cookies and candy bars. The salty stuff includes potato chips, buttered popcorn, nuts, and so on. I remember somebody talking about the basic food groups: salt, sugar, caffeine, grease, and alcohol. Maybe THOSE are the types of junk food.

How does junk food compare to other food?

It costs a lot for not much food value. Usually it only tastes good. It doesn't even leave you feeling good for long, no, not even chocolate. "Real" food, on the other hand, may or may not taste good (the stuff your mother makes you eat), but usually there is some genuine nutritional value to it, like in a big salad or a baked chicken, which I love. I can feel pleased with myself for eating properly when I eat chicken and salad, instead of feeling guilty after garbaging down on a candy bar or a gallon of popcorn.

What is the present status of junk food?

PROFITABLE!

How should junk food be interpreted?

As an indication of how wealthy our society really is. People who are on the edge of survival don't eat junk food. Junk food is a product designed for people who can afford to pamper themselves a little.

What are the facts about junk food?

I need to look up some of these — like how many jillion french fries Americans eat per year, how many pounds of chocolate? How much ice cream? How many gallons of soft drinks?

How did junk food happen?

I have no idea. I know ice cream has been considered a spectacular treat for at least a couple of centuries. Roman emperors used to send to the Alps for ice to make fruit ices, so my theory that junk food is a manifestation of a society's wealth seems to hold true. Chocolate candy has always been a fancy gift, too.

What kind of person is junk food?
I think I can omit this question.

What is my personal response to junk food?
I love the stuff but don't dare eat much of it because it appears immediately — as in overnight — on my rear end. I like cola drinks and chocolate stuff and french fries as much as anybody. I have to leave it all ALONE, though, because I can't eat just one. It is total abstinence for me or binge city.

What is my memory of junk food?
The first time I ever ate a hot fudge sundae I thought I'd died and gone to heaven. I get about one a year, now. My grandfather had a drugstore, and he used to make me rootbeer floats on hot summer days. My mom never keeps much junk around, but my dad buys chips and candy bars for himself a lot.

What is the value of junk food?
Like I said, making people happy even if they are getting fat and clogging their arteries while they do it.

How can junk food be summarized?
It is fun, it is nutritional garbage, it is an indicator of wealth.

What case can be made for or against junk food?
The case for it is pretty obvious. It makes people feel good by giving them the sugar, salt, fat, caffeine, and other anti-nutrition they want. Eating a whole pint of ice cream is the next best thing to having your mother kiss you and make everything better. The case against junk food is strong, too. It is bad for you in all sorts of ways. But eliminating it seems to be out of the question. Millions of potato chip junkies would rebel. And half the chemists in America would be thrown out of work.

Comments

This student has repeated herself a little, but the redundancies have helped her identify a couple of ideas she consistently brings up: junk food as a source of pleasure and junk food as an indicator of a society's wealth. Either of these ideas could serve as a thesis idea for a larger paper.

USE THE TWENTY QUESTIONS TO DEVELOP A THESIS IDEA.

The writer has also used the questions to identify what she needs to research further. She finds that she has a fair amount of general information on junk food, but no specific facts. A trip to the library is clearly in order, but she will go armed with a set of specific questions that an almanac or other sources can answer fairly quickly.

ASSIGNMENT

Monitor your spending habits for a week. What do you spend the most money on to buy regularly? Eliminate necessities such as regular meals or gasoline for your car. Identify an item you could do without but clearly enjoy because you spend money on it frequently. Examples could include movies, cable TV service, junk food, romance novels, music.

Apply the Twenty Questions to the item you selected. Look for a thesis statement in your invention work.

Write an essay in which you explore the significance of the item you chose, either for you personally or in a larger context. Be sure to gather the facts you need.

Example

WHAT JUNK FOOD REALLY MEANS

Why do millions of people who know better stuff their faces with tons of grease, salt, sugar, caffeine, and other not-so-healthy substances every year? When our supermarkets are full of the freshest vegetables and fruits and inspected, trimmed meats, why do we buy Twinkies and HoHo's? Why is it that even though our advertising suggests that we should be slender, Americans eat 17 pounds of ice cream apiece every year? Obviously, we do it because this stuff tastes wonderful and makes us happy while we eat it. However, there is a less obvious reason that may be more important. Junk food is a luxury, and the amount of money we spend on it indicates

how wealthy Americans really are in comparison with the rest
of the world.

Junk food is big business in America. For example, two of
our biggest companies compete to sell the most cola drinks.
Just recently, the biggest corporate takeover in history involved
two companies that sell a lot of junk foods, RJR and Nabisco.
Sometimes these companies create markets for products that
nobody wanted before they were invented. Do we really need
more creme-filled cupcakes or salted munchies, after all? And
yet we keep buying them.

Americans know that junk food is not very good for them.
We hear about food colorings that have to be banned because
they cause cancer. We know that cottage cheese has half the
calories and five times the protein as a piece of pecan pie,
but we prefer the pie. Partly it's because the pie just tastes
wonderful. And we've all heard about how chocolate makes
your brain think you're in love, which explains why we eat
a lot of it. Junk food is fun.

Eating junk food satisfies all kinds of cravings, from the
comfort of a cold rootbeer float on a hot summer day to
eating a box of chocolate candy after a visit to the dentist,
just to get even. I think we eat a lot of it just to show our
mothers who is really in control of this food business. She can
send all that spinach to the starving children in India, and
you can live happily on burgers, fries, and shakes. It satisfies
our senses, too. Chocolate candy doesn't just taste good, it
smells good, and if it has nuts in it, the crunchiness is
satisfying.

We associate junk food with good times. We serve our
friends pretzels and potato chips and cokes when they come
over. We eat ice cream and peanuts at the baseball game.
We meet our friends at the fast food joint and have a coke
and some fries. We binge on ice cream or oreos to get over
a bad time.

We have created a system in which we use junk food to
signal a happy occasion to each other, parties and games. The
system is supported by huge corporations that spend and make
millions of dollars on these products. Yet you don't hear
about Polish junk food or Gambian junk food or Bolivian
junk food. Is junk food only American?

I don't think so. The Swiss make great chocolate, too.
Still, most countries want to import our junk food. I think
the main reason is that they know that an economy that

includes junk food is one that has a lot of people with extra money to spend. I don't need junk food, but I spend about $11 a week on the stuff. On my budget, that is a lot of money, and I think I'm pretty typical. I may think I'm on a tight student budget, but the fact is, I have that $11 week to spend on junk. In some parts of the world, $11 is enough to feed a whole family for a week, no junk included. Even though we have a lot of problems here, the fact that we can have a segment of our economy devoted to junk food, a segment that involves incredible amounts of sugar, dollar bills, chemical compounds, chocolate, and salt, indicates that America is a very wealthy nation. Our values may be screwy, but we seem to be able to afford them.

—Linda Smith

Comments

Linda has used the thesis idea she generated with the Twenty Questions, and she has transferred some of the other ideas into her draft also. Notice, though, that in part the Twenty Questions served to provoke more thought rather than to provide specific ideas. The playful tone of the invention effort still occurs in the essay, but the essay as a whole has a slightly more serious approach to the subject than the invention originally had.

Linda went to the library and found some specifics to support her ideas, such as the number of pounds of ice cream consumed by Americans each year. Knowing that she needed some facts like these made her search easy. She found everything she wanted in a single source, an almanac. These facts strengthen her paper. She doesn't stop to define junk food, but her comparison between cottage cheese and pecan pie does the job by implication: junk food has a lot of calories for little nutritional value.

The topics and the Twenty Questions are good, nuts-and-bolts invention systems. While they are not easy to memorize, they are easy to use. They will serve you for almost any writing project. If you choose to master only a few systems of invention, try these two "Old Faithfuls."

Chapter 6

Once you have mastered the basic invention systems, you may want to try some that require learning new concepts. These really aren't hard; they just have some terminology that may look unfamiliar to you. Don't worry about it, though. Remember that **any results are good results in invention**.

The same principles apply to these systems as you have used in all the others:

Write as much as you can.
Keep at it.
Forget about editing.

Burke's Pentad

The terms Kenneth Burke uses to describe his system may sound like Greek to you; well, they **are**. Pentad comes from the Greek word for five. There are five parts to the Pentad.*

At first examination, the Pentad resembles the journalistic formula of who, what, where, when, why, and how. However, Burke's system is more complex.

First, Burke distinguishes between *action* and *motion*. An action is driven by motive. It occurs as a result of will. A motion, on the other hand, simply happens, irrespective of will. Sneezing and shivering in the cold are motions. Making

*Burke, Kenneth. *A Grammar of Motives*. Englewood Cliffs, NJ: Prentice-Hall, 1945.

an appointment with the doctor and putting on a heavier coat are actions. Burke makes this distinction because his purpose is to analyze human action. While the Pentad and the classical topics both provide analytical tools, Burke's system is designed specifically to analyze why human beings do things.

Second, while using the Pentad requires you to identify each of the five parts, its real significance lies in determining the *ratios* between any two parts. Burke does not refer to specific numerical ratios, of course; he refers instead to approximate degrees.

Determining these ratios leads you to understand cause and effect relationships. While the classical topics include cause and effect, the topics do not provide for as thorough an investigation of this relationship as the Pentad. When your subject involves the actions of people, the Pentad is especially helpful.

The purpose of the Pentad is to explore why people do things. When you are asked to determine why a character in a book does something or if you are supposed to analyze an author's purposes, use the Pentad.

USES FOR THE PENTAD IN LITERATURE CLASSES

Explaining why characters do things

Explaining why authors do things

The Five Categories of the Pentad

Act

What was done? The action must have human will behind it.

Agent

Who did it? Specifically, which person or persons performed the act?

Agency

How was it done? With what means? Was the agency a physical object, such as a weapon, a tool, or a machine? Was the agency language, such as a threat, a warning, a promise, an explanation, a metaphor, a compliment? Was the agency an argument or line of reasoning? Was the agency an organization or a person?

Scene

When was it done? Where was it done? The time and place may be specific or may represent a complex of ideas, events, and persons, such as the Renaissance or the Reign of Terror.

Purpose

Why was it done? Announced purpose? Real purpose?

Each of these categories may be combined with any of the other four to determine a ratio. Twenty combinations are possible.

Act : agent	Agent : act	Agency : act
Act : agency	Agent : agency	Agency : agent
Act : scene	Agent : scene	Agency : scene
Act : purpose	Agent : purpose	Agency : purpose

Scene : act	Purpose : act
Scene : agent	Purpose : agent
Scene : agency	Purpose : agency
Scene : purpose	Purpose : scene

To respond to all twenty combinations will generate a large quantity of material; much of it will be redundant. On the other hand, if you respond carefully to at least five or six of the combinations, you should be able to detect which ratios will be significant, so you can work more specifically with those.

EXAMPLE

Act: In <u>Gone with the Wind</u>, Scarlett O'Hara shoots and kills a Union deserter.

Agent: Scarlett

Agency: Her dead husband's pistol

Scene: During the Civil War, in the O'Hara plantation house. This action occurs after Sherman's march through Georgia, so the area has been devastated by war.

Purpose: To defend the household against the deserter; only women, children, and a broken old man are present in the house. Also, Scarlett revenges her mother's death and the loss of her own innocence on the deserter.

Ratios: In this case the ratio of Scene:Agent seems especially significant. The effects of the Civil War on Scarlett form the plot of this novel. The war destroys the culture Scarlett was reared to enjoy; instead of living as a pampered, wealthy belle, Scarlett must struggle simply to avoid starvation. She supports herself and a family. The war changes her values also. Once jealous of her sisters and loathing of her sister-in-law Melanie, Scarlett defends them against the deserter. The violence of the war, which she has witnessed repeatedly, becomes part of her; she becomes capable of murder.

The ratio of Agency:Agent is also significant. Scarlett's husband, Charles Hamilton, dies before ever encountering combat. His belief that Scarlett is brave and loving annoyed her during their brief marriage, and she is embarrassed that he failed to be heroic. Thus it is ironic that she uses his pistol to kill the enemy.

The Agent:Act ratio is important because the book develops the theme that women are capable. Scarlett consistently engages in activities her neighbors consider unfeminine. Using the pistol to defend her home against an enemy is an example of Scarlett's actions involving rejection of traditional gender roles.

Comments

This writer is beginning to develop some interesting ideas about Scarlett O'Hara. A thesis statement for a character analysis seems to be lurking in the last segment, something about Scarlett's struggle against traditional ideas about femininity.

The student can write about Margaret Mitchell's techniques, such as irony. Or the student can write about Scarlett herself, pointing out that most of her actions are motivated by rebellion against traditional ideas about women.

ASSIGNMENT

Apply the Pentad to a character in a short story, novel, play, or poem that you have read recently.

Identify the Act, Agent, Agency, Scene, and Purpose you will discuss.

Set up at least three ratios and discuss them.

ASSIGNMENT

If you wrote about a character in a short story or novel, use your invention to develop an essay in which you discuss which ratio you consider most important for the character.

If you wrote about a character in a play, assume that you are to act that role. Write a report to the director of the play, explaining the motives of your character.

If you wrote about a character in a poem, explain in an essay what information in the poem (words, phrases, allusions) led you to set up the ratios you selected.

EXAMPLE

The example for this assignment is a response to Theodore Roethke's poem, "My Papa's Waltz," about a son's remembrance of a childhood bedtime romp with his father.

THE MOTHER IN "MY PAPA'S WALTZ"

Act: We don't see the mother act. Maybe nonaction is the point here. The important thing is that the mother is not part of the action between the boy and his father.

Agent: The mother. Whose "countenance could not unfrown itself."

Agency: Her face. Roethke can't call it a face, though, because face is too friendly a word. He uses "countenance" instead. The word face also is a verb, meaning to really see and really accept like in facing the facts. The mother doesn't face and doesn't have a face. She has a countenance that can't unfrown itself. A frozen expression.

Scene: In her own home, among her family. It is more important that she is so frozen here than it would be outside.

Purpose: Her purpose is to disapprove. That is why Roethke doesn't call her "Mama" to go with my "papa." He calls her "my mother" which suggests disapproval. So does that frown-frozen countenance. So Roethke's purpose is to tell us how far from this warm play the mother really is.

First Ratio The Agency/Purpose ratio. The mother's face and her disapproval work together. The purpose of disapproval makes the face frozen.

Second Ratio From the author's point of view, Agency/Purpose shows up when Roethke wants to tell us about this character. His purpose is to show what she is like. So he uses words like "countenance" and "my mother" to get the idea across. The words are the agency.

Third Ratio Agent/Scene ratio. The mother is at home with her family at bedtime, which should be a really pleasant occasion for fun and love. But she is only there physically. She is not "there" with the father and son. She doesn't dance or laugh. She stands apart disapproving. Roethke uses this contrast between what is and what ought to be to tell us a lot about the family.

ASSIGNMENT

Pick a work you have had some difficulty understanding. Identify the place where you stop being able to keep up with the work. This point could be a word or a sentence, a chapter, stanza, or scene.

Use the Pentad to clarify your information about the work at this point of mystery. Identify an act, agent, agency, purpose, and scene, either for the characters in the work or for the author. Try to set up at least three ratios. Keep working until you develop a set of ideas that seems to explain the work and that solves your problem.

What if I'm not writing about literature?

Another invention system that helps student writers was designed to encourage you to look at an event from several points of view. Often, we forget that others don't necessarily see the same details we do. What you think is important in an event depends in part on your role in the event. **Prism thinking** addresses these differences.

Prism thinking works especially well for reporting events and for analyzing their meaning. History papers are a perfect place to use prism thinking, as are sociology papers and papers for the social sciences.

Prism thinking

A short, easily remembered system developed by Maxine Hairston, prism thinking resembles the Pentad's approach to a subject.* Like the Pentad, prism thinking concentrates on human action.

In prism thinking, writers examine a subject from three different points of view. Hairston points out that it is not always possible to approach a subject from all three points of view. You can decide if that is the case only if you **try** all three points of view, though.

---

*Hairston, Maxine. Contemporary Composition. 4th ed. Boston: Houghton-Mifflin, 1986, p. 50.

Participant

This point of view requires you to discuss events or a subject in which you play an active role. The participant view resembles first-person narration in fiction.

Spectator

The spectator role requires you to discuss events from the point of view of a participant other than yourself. The spectator view is like a limited third-person narration in fiction.

Reporter

As a reporter, you discuss events as a disinterested observer. This view corresponds to third-person, omniscient narration in fiction.

EXAMPLES OF PRISM THINKING

Suppose you have chosen to write about a rock concert at the State Fair; the headliner band was from Australia.

PARTICIPANT
The opening band was better than most people realized. They were too anxious for the headliner group.

I could see really well; my tickets were the best I've ever had for a rock concert. I was on the fourth row, and the girl in front of me was short.

This is my favorite band. I have all of their albums. Their live music is different from their recorded music, stripped down and simpler. I can't tell if I like it better than the polished, carefully mixed studio sound. I do know that I really liked what I heard. They seemed to really enjoy playing with each other. The improvisations and long solos would never happen on an album. I'm really glad I got a chance to see that their talent isn't created by an electronic wizard.

SPECTATOR
The physical arrangements for a concert in the States are less intimate than in Australia. The members of the band

*have to communicate with the audience over a large empty
expanse — there is no chance for hysterical girls to grab at the
lead guitarist, no chance to see anybody dance.*

*The band is relieved that the sound system is holding up
in spite of being loaded and unloaded into trucks three times
in as many days.*

*It is not as hot here as in Australia, even though the
newspapers say the state is enduring a record heat wave.*

*The band's Australian slang goes right by these audiences;
they don't understand Australian humor, much. But they like
Australian rock a lot.*

*The band wonders where American kids get the money to
buy $15 tee-shirts on top of $18 concert tickets.*

REPORTER

*The concert was held on the fifth night of the State Fair,
a night when the teenagers who can drive here do. The
headliners on other nights appeal to the older crowd, but this
one was strictly for people under 25, though one or two
ancients of 40 appeared in the audience.*

*The band is making a two-month tour of the U.S. They
have never played the State Fair circuit before, usually doing
college campuses during their American tours.*

*The concert was a sellout weeks before the Fair even
started. Fair officials estimate that they could have sold out a
second concert if it had been possible to schedule one.*

*The crowd was a well-behaved one. Fair security police
arrested a couple of people for possession of marijuana;
otherwise there were no incidents.*

Comments

The exploration of this subject through prism thinking has
generated several possible approaches. The band viewed the rock
concert audience as apparently young, well-behaved, and afflu-
ent. The writer has also discovered some significant differences
between live and recorded music and has had an insight into
the talents of the Australian band. The fifth night of the State
Fair seems to be a summer tradition for area teenagers. Any of
these observations could provide a focus for the essay.

The spectator focus has adopted the band members' point of
view. Here, the ratio between agent and scene, to use Burke's
terms, is different from the ratio indicated in the participant

section. The band member sees considerable significance in scene. The participant, on the other hand, sees act and agency as the most important ratio; the concert music of the band is different from its studio music.

If you must perform an exhaustive analysis of events in order to understand the interactions of people, combining prism thinking with the Pentad will yield a large quantity of useful information and perhaps provide the insight you need. Each point of view will imply different ratios; a participant may see the connection between agency and setting differently from the way a reporter would see it.

You can do a simpler combination that still will yield a great deal of information if you use prism thinking in combination with the Journalist's Questions.

ASSIGNMENT

Using the Journalist's Questions, collect information about an incident in which you were treated badly or unjustly. Use the questions to get the specific facts about time, place, participants, and so on.

Use prism thinking to describe the incident three different ways.

Has this exercise changed the way you view the incident? If so, how? Was the injustice more or less serious than you initially thought?

Try using the Pentad to examine the motives of the person or persons who caused the injustice.

Chapter 7

What if I want to learn something more challenging?

If you want a black belt in invention, there is one more system you will want to study. It is called **Tagmemic Invention**. If you want to know how this system came to be called tagmemic, read the next section. Otherwise, you can skip directly to an explanation of how to use it and to the example.

Tagmemic Invention is very powerful; it is a thorough system for examining any topic, and after you have used it, you should understand what you are writing about very well indeed. It isn't an easy system to memorize. Probably you will have to have the system in front of you whenever you want to use it. However, for examining any subject, say for a term paper or any lengthy project, Tagmemic Invention is an effective method.

Why is it called Tagmemic Invention?

Tagmemic Invention derives its name from a term used in linguistics. A tagmeme is a grammatical unit; its definition parallels the structure of the system that follows, including the unit's unique features, its range of variation, and its relationship to other grammatical units in a system. The developers of Tagmemic Invention, Young, Becker, and Pike, understood that their method for analyzing the tagmemes in an ancient language could be applied to other subjects, that the method is useful for more than just linguistic studies.* Its terminology is borrowed from another discipline, physics.

*Young, Richard, Alton L. Becker, and Kenneth L. Pike. *Rhetoric: Discovery and Change*. New York: Harcourt, 1970.

How does Tagmemic Invention work?

This system works by taking you through a series of guided steps. These steps make you see the subject you are examining from a variety of points of view. In that respect, it is like prism thinking. However, prism thinking makes you change who you are as you look at something, while Tagmemic Invention makes you change the way you look at the subject itself.

The unique advantage of Tagmemic Invention is that it examines your subject as a changing, dynamic thing. The other systems you have tried tend to emphasize the subject you write about as a static, completed matter. Change, however, is a significant issue in some subjects.

In addition, Tagmemic Invention helps you to look at your subject both separated from and included in larger systems, like taking a gem out of its setting and then putting it back in. The gem looks different, depending on how you examine it. Your subject will do the same.

Tagmemic Invention requires that you look at your subject as a "particle" or a single, isolated unit. Then you look at it as a "wave" or a unit of coherent movement or change. Then you look at it as a "field" or as a member of a category.

The originators of Tagmemic Invention first presented it in the form of a matrix or grid, with nine steps corresponding to the combination of three items on the vertical axis and three items on the horizontal axis. The steps are presented both in the grid and in a straight sequence here, however, because the first time you use them, you may need a little extra explanation.

The system is designed to help you look at the "big picture," a global view of your subject, and then at the "little picture," a worm's-eye view. The matrix also encourages you to examine change from both points of view. Most invention systems operate with a kind of stop-action approach; you look at your subject as it is frozen for the moment, like a still photo. Tagmemic Invention keeps the film rolling; change is inherent in most matters, and this system accounts for change, resembling a short movie instead of a still shot.

In Tagmemic Invention, as in all the other forms of invention, expect redundancy, expect the unexpected. System D is in full swing here — whatever happens is okay. Don't worry if you are doing it "right." Just keep trying it. You'll come to like this system more and more as you use it.

TAGMEMIC INVENTION

1. **View the unit as an isolated, static entity.** *What are its contrastive features; that is, the features that differentiate it from similar things and serve to identify it?*

   This step is essentially the same as definition from the classical topics.

2. **View the unit as a dynamic object or event.** *What physical features distinguish it from similar objects or events? In particular, what is its nucleus?*

   Look at your subject as a sequence of motions or events. Where does the sequence begin? Where does it end? What helps you to determine which events or motions are part of the sequence? How do you tell which ones are not part of the sequence?

3. **View the unit as an abstract, multidimensional system.** *How are the components organized in relation to each other? More specifically, how are they related by class, in class systems, in temporal sequence, and in space?*

   Examine how the parts of your subject relate to each other. Are there categories of parts? What are they? Is there a series of sequences within your subject?

4. **View the unit as a specific variant form of the concept; that is, as one among a group of instances that illustrate the concept.** *What is the range of physical variation of the concept, that is, how can instances vary without becoming something else?*

Your subject is probably an example of something. What is it an example of? What are the characteristics that you **must** have to identify an example? What characteristics cannot be removed? What characteristics can be changed? How much?

5. **View the unit as a dynamic process.** *How is it changing?*

   Most things undergo change in time. Look at your subject and discuss how it is changing from the way it was in the beginning.

6. **View the unit as a multidimensional physical system.** *How do particular instances of the system vary?*

   Go back to the parts of your subject that you examined in item 3. What kind of variation in detail do you find among the parts?

7. **View the unit as part of a larger context.** *How is it appropriately or typically classified? What is its typical position in a temporal sequence? In space, that is, in a scene or geographical array? In a system of classes?*

   Where does your subject belong? If you were going to find it in the library, what categories would you consider? Does your subject seem to belong to a particular set of circumstances? A particular place or time? A particular group of things?

8. **View the unit as part of a larger, dynamic context.** *How does it interact with and merge into its environment? Are its borders clear-cut or indeterminate?*

   What things outside of your subject affect it? What things does it affect? Do any of these things significantly change the whole system?

9. **View the unit as an abstract system within a larger system.** *What is its position in the larger system? What systemic features and components make it part of the larger system?*

Where does your subject fit into a larger scheme of things?
What part does your subject play? How does it do that?

EXAMPLE

The following example presents responses to the Tagmemic
Invention system in sequence. Each section of the example
is numbered to correspond to a section of the tagmemic
system.

VIEW THE UNIT AS AN ISOLATED, STATIC ENTITY
1. Lewis Carroll's <u>Alice's Adventures in Wonderland</u> is a
   children's fantasy novel written in 1863. The revised
   version illustrated by John Tenniel was published in 1865.

VIEW THE UNIT AS A DYNAMIC OBJECT OR EVENT
2. <u>Alice</u> differs from other Victorian children's literature
   because it makes no attempt at didacticism or moralizing.
   In fact, the book satirizes the moralizing of contemporary
   works.

VIEW THE UNIT AS AN ABSTRACT, MULTIDIMENSIONAL SYSTEM
3. The book is made up of episodes related primarily by the
   presence of the character Alice. Certain motifs unify the
   book, such as eating or drinking in order to adjust Alice's
   size and using puns and word games. Alice sips from a
   bottle labeled "drink me" and becomes so small that she
   nearly drowns in the pool of tears. Later, she drinks
   something at the White Rabbit's cottage and becomes too
   large to fit in the house. The Caterpillar's mushroom also
   adjusts her size. Word games include Carroll's revision of
   Watt's poem "How doth the busy little bee" into "how
   doth the little crocodile," Carroll's punning chapter title
   about the White Rabbit sending "a little Bill," and the
   Mock Turtle's name.

VIEW THE UNIT AS A SPECIFIC VARIANT FORM OF THE CONCEPT, THAT IS, AS ONE AMONG A GROUP OF INSTANCES THAT ILLUSTRATE THE CONCEPT
4. <u>Alice</u> has been retold in shortened versions, in versions

*illustrated by other artists, in musical stage versions, and
in a Walt Disney cartoon feature. Many of these versions
borrow the episodes from Carroll's book but substantially
alter the original qualities of the book. For example, in
some versions Alice's independent, somewhat quarrelsome
personality is replaced by a saccharine sweetness, even
helplessness. In the Disney film, Alice weeps helplessly in
a dark wood, wanting to go home. She is rescued by a
creature that never appears in Carroll's story. This
sequence echoes The Wizard of Oz more than Alice.
Carroll's book is about manners, yet many later versions
focus instead on magic, on Alice's disappearance down the
rabbit hole, or on her struggle to return to reality. Carroll
concentrates on Alice's efforts to impose the system of
behavior she has learned from her adult supervisors on the
inhabitants of Wonderland. This theme of the tension
between childish impulses and adult requirements is a
major one in the book, particularly in the Mad Tea Party.
Many retellings avoid this theme, missing the central
appeal of Alice to children.*

VIEW THE UNIT AS A DYNAMIC PROCESS

5. *Alice becomes more and more self-sufficient as the story
   progresses. At first she is deferential, hesitant, careful. As
   she meets characters who do not subscribe to her code of
   behavior, she questions them and herself. The arbitrariness
   of some rules and the logic of others becomes apparent.
   Alice's Adventures in Wonderland is sometimes
   condemned as inappropriate for modern children. Its
   distinctly Victorian references and style have been
   challenged as inaccessible to modern children. Many of its
   most devoted readers are adults. Some child development
   experts argue that its violence and anti-adult stance make
   it a poor choice for children. However, others argue that
   its themes appeal to children, that its wordplay suggests
   language as a fun thing, that the violence is couched in
   fantasy and is therefore inoffensive. Others argue that
   shielding children from violence is impossible and that the
   violence in Alice is similar to the kinds of violence
   children invent and inflict anyway. This controversy is
   relatively new; until recently, Alice enjoyed an undisputed
   reputation as a children's classic.*

VIEW THE UNIT AS A MULTIDIMENSIONAL PHYSICAL SYSTEM

6. At first reading, the characters in <u>Alice</u> could be classified as friendly to Alice or unfriendly to her. However, a more useful classification would divide them between those whom she finds interesting, those whom she finds annoying, and those whom she finds threatening. Thus she is attracted to the White Rabbit, the Duchess, and the Cheshire Cat. She is annoyed by the Caterpillar and the Mad Hatter. She is threatened by the Queen of Hearts. Alice edits her conversation with these characters, based on which category they fit into, a further illustration of Carroll's theme of manners.

VIEW THE UNIT AS PART OF A LARGER CONTEXT

7. <u>Alice</u>'s importance lies in its rejection of traditional themes and formulas for children's literature. It denies the effectiveness of moralizing, preachy content. In an age of the careful moral content of books such as <u>King of the Golden River</u> or <u>Little Women</u>, the appearance of <u>Alice</u> must have startled many. Yet it enjoyed immediate success.

VIEW THE UNIT AS PART OF A LARGER, DYNAMIC CONTEXT

8. Persons unfamiliar with the complexity and richness of children's literature often assume that children need gentle, didactic works that encourage "good" behavior. <u>Alice</u>'s significance lies in its rejection of this theory. Instead, Carroll accepts children as they are and addresses the issues of language and behavior which interest them. He addresses his audience directly rather than through their parents or their teachers.

VIEW THE UNIT AS AN ABSTRACT SYSTEM WITHIN A LARGER SYSTEM

9. <u>Alice</u> is one of the earliest examples of a children's work that abandons preachiness or didacticism. In fact, Carroll reverses the usual methods of didacticism by having the child in the story look for the morals of the stories she hears and having her attempt to impose mannerly behavior on other characters. To emphasize the point, the moral of each story is absurd or wryly realistic. Carroll tells the

truth to his readers — the world belongs less to busy little bees and more to crocodiles, which eat fish.

Comments

By working through the system, this writer has begun to develop several possible ideas for an essay about *Alice's Adventures in Wonderland*. The essay could be about how Carroll chooses not to be preachy in his book. Or the essay could be about how much the retellings of the story have changed its original meanings.

The system has also helped the writer to see Carroll's book as something besides a fantasy novel. Tagmemic Invention works particularly well for finding new ways to approach a subject, precisely because it leads you through a variety of points of view. When you need to develop a fresh approach to a much-covered subject, Tagmemic Invention should do the job.

The tagmemic system's greatest power is in forcing the writer to look at a subject in several ways, such as placing it into a larger context or dividing it into parts and examining how they fit together. The system also requires the writer to look at the issue of change, something none of the other systems requires specifically.

ASSIGNMENT

Try Tagmemic Invention on a really shopworn topic, such as "How I Spent My Summer Vacation" or "Dorm Food" or "Why I Came to This College." Underline the ideas that seem new or surprising to you. Note which sections of the system generated these ideas.

ASSIGNMENT

Apply the tagmemic system to a topic, concept, work, or issue that you have had difficulty understanding in one of your classes.

The preceding example is a response to this assignment, written by a student in a children's literature class.

THE TAGMEMIC INVENTION SYSTEM

	Contrast	Variation	Distribution
PARTICLE	1) View the unit as an isolated, static entity. What are its contrastive features, i.e., the features that differentiate it from similar things and serve to identify it?	4) View the unit as a specific variant form of the concept, i.e., as one among a group of instances that illustrate the concept. What is the *range* of physical variation of the concept, i.e., how can instances vary without becoming something else?	7) View the unit as part of a larger context. How is it appropriately or typically classified? What is its typical position in a temporal sequence? In space, i.e., in a scene or geographical array? In a system of classes?
WAVE	2) View the unit as a dynamic object or event. What physical features distinguish it from similar objects or events? In particular, what is its nucleus?	5) View the unit as a dynamic process. How is it changing?	8) View the unit as a part of a larger, dynamic context. How does it interact with and merge into its environment? Are its borders clear-cut or indeterminate?
FIELD	3) View the unit as an abstract, multidimensional system. How are the components organized in relation to one another? More specifically, how are they related by class, in class systems, in temporal sequence, and in space?	6) View the unit as a multidimensional physical system. How do particular instances of the system vary?	9) View the unit as an abstract system within a larger system. What is its position in the larger system? What systemic features and components make it a part of the larger system?

From Young, Richard, Alton L. Becker and Kenneth L. Pike. *Rhetoric: Discovery and Change*. New York: Harcourt, 1970, p. 127.

USE TAGMEMIC INVENTION FOR

Analyzing the process of change
Looking at "big picture" elements
Looking at "little picture" elements
Finding a fresh way to write on a subject.

What if I am writing a term paper?

Invention is a godsend for students stuck with the chore of developing a term paper. Just finding a workable topic is often the hardest part.

> To find term paper topics, refer to the Twenty
> Questions, prism thinking, brainstorming,
> and the Journalist's Questions.

Use the Twenty Questions and the Journalist's Questions to direct your research. These two systems will help you make sure you have all the information you need. At the same time, using them will help you decide what information to look for. Rather than just rummaging through the library for something about your topic, you will be looking for specifics, such as time, place, names, and so on. The librarian will be able to help you more efficiently if you ask a specific question, such as the date of a scientific discovery or the method for manufacturing a particular product.

Be sure to check your authorities carefully as you do your research. **Review the section on authority.**

Use looping, prism thinking, and Burke's Pentad to develop a thesis. Remember the center of gravity sentences in looping? Try finding a center of gravity sentence after you do some prism thinking. If your paper topic involves the actions of people, the Pentad can be a helpful method for finding a thesis idea.

Finally, the following are two topics left over from the classical system of topics. These two are fairly specialized, but they are especially suited to the demands of term papers.

STATISTICS

Polls, opinion surveys, bestseller lists, and television ratings are statistical authorities writers frequently use. Statistics may tell you what people think; a carefully constructed survey can tell you why they think it.

Mark Twain described the three kinds of lies as "Lies, damn lies, and statistics."

A good writer will use statistics with some caution even though they often make compelling evidence to support an argument. For example, a poll may say that 60 percent of users prefer a particular brand of a product. Sixty percent, however, could represent thousands of people or only three or four, depending on the size of the population surveyed.

If you use statistics to support an argument or to add to your information on a subject, you must examine them carefully. The statistics should actually support any inferences drawn from them.

> What is the source of the statistics?
> Has the source produced reliable information in the past?
> Does the source use appropriate methods for collecting statistics?
> How large was the sample studied?
> Was the sample representative?
> How recent is the information?
> Do any studies or polls exist that contradict or alter the conclusions drawn from the statistics?

Be sure to answer these questions if you use statistical information in your term paper.

LAW

This authority gains its power from respect for the written word. The topic of law includes documents such as wills, statutes, contracts, certificates, and other written records. In argumentation, evidence such as letters, birth certificates, wills, contracts, or court decisions carries substantial weight.

The strength of these pieces of evidence derives from their concreteness. They can be touched, pointed to, read and reread. While Justice John Marshall is long dead, his opinions are still available for reading and use in court. An entry in a parish record will tell a genealogical researcher that an ancestor was born on August 14, 1703. A passenger list filed with the White Star Line confirms exactly who was on the *Titanic*.

Because this kind of evidence is so concrete, it is easy to accept it without questioning its validity or accuracy. As you accumulate information under the topic of law, you should examine the quality of your information carefully.

What is the source of the information?
Is the source a reliable one?
Does any other documented evidence contradict or supersede this evidence?

Remember that signatures can be forged, dates misremembered. For example, historians have long questioned the authenticity of the Casket Letters, documents that seem to confirm Mary Queen of Scots' involvement in a plot against Elizabeth I. The letters may be regarded either as evidence that Mary was indeed guilty or as evidence that someone wanted her to be convicted. Similarly, establishing the exact dates of events is not always simple. The date found by that genealogical researcher may refer to the date of birth or it may refer to the date of baptism or it may simply be the date the parish clerk decided to enter.

The preceding topics are often helpful when you must develop arguments to support your position. They are also particularly applicable to term paper writing.

Use invention to solve these problems in term paper writing:

- Finding a topic
- Conducting directed research
- Validating the quality of your sources
- Developing a thesis idea
- Writing a paper that contains strong, carefully presented information

Is there anything else I need to know about invention?

If you have learned all the systems in this book and if you have tested all of them on real assignments, then you should be very comfortable with invention by now. The more you use these systems, the stronger they become and the stronger **you** will become as a writer.

There is a fancy name for these systems: heuristics. A heuristic procedure is a systematic exploration of a subject, with no specific outcome predicted. If you are going for that black belt, you should start talking about **heuristic procedures**, which are what you have used all through this book.

Eventually, if you use these heuristics enough, you will begin to find others. You will learn to recognize them. If you have had success with these, maybe you'll try the new ones when you find them. Just look for systems of questions or investigations.

These invention systems are intended to be more or less universal in their application. There are suggested uses for each one, but they are only suggestions. You can use any system any time, any way that seems helpful.

One Last System

Any invention system is going to be more effective if you know why you are using it. All writers write in response to some need or problem. If you identify the purpose of your writing, you can direct your invention toward that purpose. The following is a short system of identifying the kinds of problems that generate most writing, both in school and out in the working world.

PROBLEMS OF INFORMATION: Do you have the information you need? What information do you want? [Journalist's Questions; Twenty Questions; classical topics].

PROBLEMS OF COMMUNICATION: Can you get the information distributed? Are there obstacles to communication? [classical topics; Pentad; Tagmemic Invention].

PROBLEMS OF PERFORMANCE: Can a task be done? How? By whom? What obstacles exist to getting the job done? [Journalist's Questions; freewriting; Tagmemic Invention; looping; brainstorming; clustering].

PROBLEMS OF POWER AND CONTROL: Who has power? Who wants power? How are power issues involved in information, communication and performance? [Tagmemic Invention; Pentad; prism thinking].

Use the invention systems you have learned to solve these and other problems.

**Above all, remember System D:
If invention seems to be working, it is.**

Index